Environmental Management and Economic Development

A World Bank Publication

Environmental Management and Economic Development

Gunter Schramm and Jeremy J. Warford
editors

Published for the World Bank
The Johns Hopkins University Press
Baltimore and London

© 1989 The International Bank
for Reconstruction and Development / THE WORLD BANK
1818 H Street, N.W., Washington, D.C. 20433, U.S.A.
All rights reserved to material not previously copyrighted.

Manufactured in the United States of America
First printing July 1989

The Johns Hopkins University Press
Baltimore, Maryland 21211, U.S.A.

Library of Congress Cataloging-in-Publication Data

Environmental management and economic development / Gunter Schramm and
    Jeremy J. Warford, editors.
        p.   cm.
    Includes bibliographies.
    ISBN 0-8018-3950-5
    ISBN 0-8018-3904-1 (pbk)
    1. Economic development—Environmental aspects.   I. Schramm
Gunter.   II. Warford, Jeremy J.
    HD75.6.E57   1989   65574
    363.7—dc20   89-8001
                                                                    CIP

Contents

Acknowledgments

The chapters by David Pearce and Anil Markandya, Norman Myers, Robert Repetto, and John A. Dixon are based on studies originally prepared for an environmental workshop sponsored jointly by the Economic Development Institute and the Environment Department of the World Bank. Earlier versions of these chapters as well as those by Jeremy J. Warford, Kenneth J. Newcombe, Dennis Anderson, and Gunter Schramm were published in a special edition of the *Annals of Regional Science*, vol. 21, no. 3, November 1987, entitled "Environmental Managements and Economic Development." They are reprinted here with permission by the *Annals*.

Contributors

Dennis Anderson, *Chief Economist, Shell Oil International, London (on secondment from the World Bank)*

Jane Armitage, *Economist, Agriculture Operations, Southern Africa Department, World Bank*

John A. Dixon, *Research Associate, East-West Center, Honolulu*

Salah El Serafy, *Adviser, Economic Advisory Staff, World Bank*

Ernst Lutz, *Economist, Economics and Policy Division, Environment Department, World Bank*

Dennis J. Mahar, *Chief, Population and Human Resources Division, East Africa Department, World Bank*

Anil Markandya, *Associate Professor, Department of Political Economy, and Associate Director, International Institute for Environment and Development, University College London*

Norman Myers, *Consultant, Environment and Development, Oxford*

Kenneth J. Newcombe, *Senior Operations Officer, Industry and Energy Operations, Eastern Africa Department, World Bank*

David Pearce, *Head, Department of Political Economy, and Director, International Institute for Environment and Development, University College London*

Robert Repetto, *Senior Economist and Program Director, World Resources Institute, Washington, D.C.*

Gunter Schramm, *Chief, Energy Development Division, Industry and Energy Department, World Bank*

Jeremy J. Warford, *Senior Advisor, Environment Department, World Bank*

1

Introduction

Gunter Schramm and Jeremy J. Warford

This volume is devoted to environmental management and economic development, with special reference to the problems of developing countries. This huge topic includes the multifaceted problems associated with the construction of hydropower dams and reservoirs; development of mines and irrigation projects; land clearing, deforestation, and overgrazing that cause erosion, siltation, and flooding; and the many health hazards of overcrowded city slums such as inadequate water supply, lack of sewerage and garbage disposal facilities, and life-threatening air pollution. Although generalizations are difficult to make, it appears that some of the most serious and pervasive environmental problems facing developing countries tend to be both a cause and effect of poverty.

Poverty, of course, has been the prevailing condition of the vast majority of people throughout the developing world for centuries. But poverty in earlier times did not inevitably cause unacceptable and lasting environmental damage, because populations, by and large, were stable and lived within the confines of their existing resource base. A hallmark of the last several decades, however, has been the rapid growth of population throughout the developing world. This growth forces more intensive use of land and water resources, and thrusts people into more marginal, environmentally sensitive lands. Population growth leads to rapid forest depletion, and it fills the cities with relentlessly growing multitudes of human beings who have few or no skills, no jobs, and no resources. Poverty is the prevailing condition, both in the cities and the countryside. And poverty compels people to extract from the ever shrinking remaining natural resource base, destroying it in the process. In fact, the major characteristic of the environmental problem in developing countries is that land degradation in its many forms presents a clear and immediate threat to the productivity of agricultural and forest resources and therefore to the economic growth of countries that largely depend on them. Those countries most affected also tend to be the poorest of the poor: the problems of the Sahelian nations are an extreme illustration of a widespread phenomenon.

The all-pervasiveness of poverty creates its own vicious cycle of destruction. Because of poverty and the growth in population, countries that rely on environmentally threatened resources are unable to shift to alternatives. For example, countries often are unable to import fertilizers because of a shortage of foreign exchange, but cow dung or crop residues, which were used traditionally, are needed as cooking fuels because forests have been cut down to make room for added agriculture or to burn for firewood. Institutional factors compound the problems. Insecure land tenure, artificially low farm prices, and lack of knowledge combine to prevent farmers from taking appropriate soil conservation measures. Urban dwellers keep using charcoal and firewood, which contributes to the drastic overcutting of forests around cities and along the few available access routes to the hinterland, and they use these fuels inefficiently because they do not know any better or do not have access to more energy-efficient cooking appliances and utensils. On balance, the problems are getting worse, and environmental stress and destruction are becoming the norm rather than exception in most of the developing world.

Nevertheless, there are grounds for at least some cautious optimism. One main reason is that, among the leadership and governments of the developing countries and the outside agencies that support them, there is growing awareness of the problems. Gradually, studies are starting to focus on these issues, policies are being examined for their impacts on environmental parameters, and some actions are being taken to prevent damage from occurring or to mitigate the consequences.

The chapters in this volume, which focus primarily on agricultural and forestry issues, consider some of these changes and emerging trends. One of the more hopeful findings is that economic development and environmental protection are not unalterably opposed to each other. On the contrary, more often than not development and protection go hand in hand. Improving one enhances the other. It is also shown that much environmental damage is the result of either lack of knowledge or shortsighted policies. Systematic evaluations and assessments quite often prove that economic betterment in the traditional benefit-cost sense can be achieved through more sensible policies and actions that protect and enhance environmental values at the same time. Environmentalists and economists, far from being natural enemies, are in fact natural allies. This is one of the major conclusions of this book.

This volume can be divided into three parts. After an overview chapter (by Warford), three contributions (by El Serafy and Lutz, Pearce and Markandya, and Myers) look at analytical and methodological questions. The balance of the chapters (by Repetto, Mahar, Newcombe, Armitage and Schramm, Anderson, and Dixon) illustrate in detail many of the problems and point to possible solutions. They are more empirically oriented than the earlier chapters, and they analyze and evaluate environmental, land, and water issues in the context of economic development objectives in various developing countries.

The Papers in Brief

Warford's overview (chapter 2) argues for the need to design broad economic policy instruments to reverse the trend in many developing countries toward increasing degradation and destruction of natural resources. It emphasizes that the natural resource base, often critical for economic development, is in many cases threatened by rapid population growth, the effect of which is compounded by inadequately controlled land and water use. Warford argues for effective policy interventions to influence the environment-related behavior of countless relatively small-scale, resource-using activities throughout a nation's economy. Management of natural resources should thus become a standard consideration in macroeconomic and sector analysis, and the physical linkages between sectors need to be critically examined. Governments must overcome major institutional and political obstacles. New approaches, which would provide incentives and rewards to policymakers, must be developed to increase interagency cooperation, avoid overlapping jurisdictions, and prevent vested interests from paralyzing these new initiatives.

El Serafy and Lutz (chapter 3) address one of the shortcomings of the conventional framework used for measuring economic growth, the system of national accounts (SNA). Within the SNA framework, gross domestic product (GDP) is usually used as the indicator of economic performance. As the authors point out, any prudent accounting of national income should reflect sustainable income as closely as possible. Under current SNA conventions, however, no account is taken of the reduction in national wealth through the depletion and degradation of natural resources. In addition, the so-called defensive expenditures incurred to protect society against unwanted environmental side effects are counted as income—that is, as increases in GDP—rather than as intermediate expenditures needed to sustain actual income. As the authors show, man-made resources are valued as productive assets and written off against the value of production as they depreciate, whereas natural resources are not: they are considered to be "free gifts." This bias provides false signals to policymakers, because it counts the nonsustainable depletion of natural wealth as income creation. The authors review a number of writings on the subject and discuss the difficulties of including environmental and natural resource effects in a national accounting system. Proposals are made on how the current shortcomings can be overcome through a modified or augmented system of national accounts.

Relationships between environmental protection, poverty alleviation, and economic growth are, of course, complex. In chapter 4, Pearce and Markandya present the principles that should be employed in estimating the consequences of resource depletion in an economy over space and time. Since renewable resources are being used up in nonsustainable ways in many countries, the costs of nonsustainability need to be enumerated and valued in order to establish the desirability of such development paths. The appropriate concept, according to the authors, is the marginal opportunity cost (MOC), a measure of the social costs of resource depletion. This concept is set in the context of models of the development process that stress the coevolutionary relationship between environment and development, rather than models that trade off material gain against environmental quality. Measures of MOC need to reflect the often intricate physical and ecological linkages within ecosystems, for example, the relationship among deforestation, soil erosion, streamflow, and sedimentation. Hence MOC combines the direct costs of resource use, the externalities arising from ecological linkages, and a user cost component that arises because of nonsustainable resource use. Formulated in this way, MOC has implications for shadow pricing exercises, national accounting, and the choice of sector and geographical areas for project appraisal.

Chapter 5, by Myers, further illustrates the complexities of the linkages between physical events and the behavior that parallels them. He stresses that economic activity depends ultimately on the environmental resources (soils, water, vegetation, climate) that underpins virtually all human endeavor. This support is obviously important for agriculture, forestry, fisheries, and hydropower generation. It is equally important for public health, because without regular supplies of good water for household use and reasonably adequate drainage the risk of disease increases. Less obviously, environmental resources have indirect linkages to other economic sectors such as communications and education. These considerations are especially significant for developing economies, which are usually more dependent on the environmental resource base than are developed economies. Furthermore, most developing countries are located in the tropics, where the resource base is more fragile and hence more susceptible to depletion than in the temperate zone. Thus, there is a premium on safeguarding the environmental resource base as an integral part of those processes that are conducive to sustainable development.

In chapter 6, Repetto illustrates the scope for policy reform in his discussion of some government policies that grossly violate both economic and ecological principles. Repetto stresses that, in order to channel development activities into sustainable patterns that preserve the productivity of natural resources, appropriate economic incentives for millions of households, farmers, and small producers in developing countries need to be implemented. Incentive problems arise both from market failures such as externalities and from policy failures such as price distortions. Policies can be improved in ways that promote resource conservation, reduce environmental damage, and simultaneously raise economic productivity, decrease government budget deficits, and ameliorate rural poverty. Revising inappropriate pricing policies for agricultural output and such purchased inputs as pesticides and fertilizers can help. Reforming the financing of public irrigation systems may improve their performance and promote better water use. Changing inappropriate revenue systems and incentives for the use of forest resources could discourage wastage.

Mahar (chapter 7) analyzes the impact of government policies on the magnitude and rate of deforestation in Brazil's Amazon region. He traces the evolution of regional development policies for Amazonia over the past twenty-five years and shows that policies and programs for road building, official settlement, and extensive livestock development have generally not been designed and carried out with respect for their environmental consequences. On the contrary, many of them have been counterproductive in economic terms and disastrous to the environment. Fiscal incentives, originally designed to encourage economic development, turned out to be intrinsically uneconomic and actually created strong incentives to destroy the productivity of land within a short time. Mahar's study provides strong evidence that many government-directed incentive policies designed to bring about development have in fact hindered it.

Newcombe (chapter 8), using Ethiopia as an example, shows in simple terms the biological links between deforestation and agricultural productivity at the subsistence level. He then quantifies the economic benefits from increased food production when firewood from rural forestry programs replaces animal dung as fuel for cooking. He shows that such strategies potentially have very high economic rates of return and provide attractive medium- to long-term benefits of added energy supplies, increased agricultural output, and environmental protection.

In chapter 9, Armitage and Schramm briefly review patterns of energy consumption in Sub-Saharan Africa. They then develop detailed strategies for continuing and improving supplies of wood for fuel, arguing that its use is not only unavoidable, but also in the best economic interest of most countries in this region. They show that intensified and more orderly utilization of wood fuels can help to enhance rather than impinge upon the environment. A case study of Malawi shows the complexities and interdependencies of the policy measures—government and private investments, regulatory and pricing initiatives, research, training and demonstration, incentives and subsidies, as well as conservation strategies—that are needed.

Anderson (chapter 10) also refers to the economic benefits of forestry projects. His case study in northern Nigeria shows that the ecological benefits, which are often disregarded in investment decisionmaking, may in fact represent the bulk of the benefits of certain types of forestry projects. If afforestation projects are analyzed only in terms of their capacity to produce wood and other tree products, their returns in arid and semi-arid areas are generally low. However, once their benefits are considered in terms of soil fertility—stemming declines in fertility and then actually enhancing it—the benefits may be appreciable. Anderson therefore argues for broadening the economic analysis of afforestation projects to include more than the traditional benefits of tree products themselves.

Finally, Dixon (chapter 11) reminds us that even at the micro level complexities abound; physical linkages, discount rates, the need for a multidisciplinary approach, and politics, all have to be considered. He notes that watersheds, as integrated land and water systems, have often proved difficult to manage successfully. This is in part due to the complexities of accommodating different groups of users and making use of multiple resources in large areas. Dixon argues, however, that watersheds are a natural physical and economic unit of analysis. As a bridge between micro-level analysis of individual farmers or economic units and a regional or sectoral approach, an integrated, multidisciplinary analysis of watershed policy yields useful results for improved management of land and water resources. Economic reasoning plays a key role in this analysis.

An integrated approach does not mean integrated implementation. Existing systems of social and government organization must be used to implement the chosen policies. This is a difficult but essential step and may require the use of government fiscal and price policies to harmonize social and private goals. Dixon's chapter, like the others in this volume, suggests that the role of the economist in this area is of critical importance, not only in carrying out traditional cost-benefit analyses, but also and perhaps even more important in devising the policy framework for more efficient environmental management.

2

Environmental Management and Economic Policy in Developing Countries

Jeremy J. Warford

The effects of poor natural resource management are being demonstrated dramatically in many developing countries. Land and water resources, which are in principle renewable, are in fact declining at rates that threaten the basis of already fragile economies (Repetto 1986b; World Commission on Environment and Development 1987; World Resources Institute and International Institute for Environment and Development 1986). The poorest countries, which tend to be heavily dependent on their natural resource base and to have relatively high rates of population growth, are the most vulnerable to the effects of environmental degradation, in part because shortages of capital and trained manpower severely limit their ability to switch to other economic activities when their natural resources can no longer sustain them. Moreover, it tends to be the poorest people in those countries who suffer most from environmental degradation.

There are, of course, innumerable ways in which the problem is manifested and generalizations are difficult to make. Air and water pollution are staggering problems in cities such as Bombay, Cairo, Lagos, Mexico City, and São Paulo. In many developing countries, however, the most critical environmental problems relate to a complex network of events: overgrazing, commercial logging and fuelwood harvesting, land clearance, deforestation, burning of crop residues and dung, soil erosion, sedimentation, flooding, and salinization. Direct economic consequences include severe reductions in energy for domestic use and in agricultural productivity, the indirect consequences of which have profound and far-reaching effects on human well-being. These prob-

This chapter is an adaptation of a paper presented at the World Conservation Strategy Conference, Ottawa, June 1986. The author is indebted to many colleagues for advice and comments, including R. Ackermann, Dennis Anderson, Norman Myers, David Pearce, Robert Repetto, Gunter Schramm, and Edward Schuh.

7

lems call for approaches that question some basic assumptions about economic development and raise generic issues about the relationships between macroeconomic planning and sectoral analysis, the handling of externalities, and the welfare of vulnerable groups and future generations.

Increasingly, the urgency of the problem is being recognized and attempts are being made to address it. Ambitious projects are under way to replace trees that have been cut down, to clear dams that have silted up because of soil erosion upstream, to pump groundwater for irrigation to help stem the advancing desert, and to clean up the poisoned ground and the polluted air. In many developing countries, however, the situation is clearly getting worse, and many of the efforts to remedy the situation are failing. In addition to lack of resources—poverty sometimes being both a cause and a consequence of environmental damage—the reasons for this include political and financial vested interests, institutional overlaps and bureaucratic inefficiencies, and the myopic view of decisionmakers. But perhaps the most important reason is the sheer difficulty of dealing with a myriad of relatively small-scale natural resource-using activities, which together are responsible for the bulk of environmental degradation.

The traditional approach to environmental management is to invest in projects that have primarily environmental objectives, such as reforestation or sewerage schemes, or to ensure that components of other projects contain elements to mitigate adverse environmental impacts. This project-by-project approach is important and must be continued. Alone it is clearly inadequate, however, and needs to be supplemented by more comprehensive, wide-ranging policies. By concentrating on curative, piecemeal solutions rather than on the underlying causes, the traditional approach—in industrialized as much as in developing countries—fails to confront the real issues, which have much more to do with the way society works than with the technical aspects of natural resource degradation. Environment-related behavior and policy are in fact at the very heart of social, macroeconomic, and sector policies—especially those relating to agriculture, energy, and industry; domestic and foreign investment; fiscal, monetary, and trade policies; income distribution; and regional planning.

It appears, therefore, that the project-by-project approach should be supplemented by one that integrates environmental and natural resource management directly into economic and social policy. This can be done in two ways: through investment programs that support environmental and natural resource objectives, and through economic, social, and institutional policies and incentives that influence the environmental-related behavior of government agencies, major resource users, and countless small-scale resource-using activities. Although our understanding of technical ameliorative measures clearly needs to be improved, the foregoing also implies a need for a greater understanding of (1) the nature, dynamics, and severity of natural resource degradation in light of economic and social criteria, including the welfare of vulnerable groups and future generations; (2) the underlying causes, both human and natural, of natural resource degradation; and (3) the range of feasible economic, social, and institutional policy interventions that are appropriate.

The Cost of Natural Resource Degradation

The National Accounts

The problem of land and water management is acute not only in ecological terms, but also in terms that economists habitually use; indeed the disciplines of economics and ecology should be seen as mutually reinforcing. In developing countries, the effects of high debt burdens and deteriorating terms of trade are compounded by the severe and escalating economic costs of natural resource degradation. Nevertheless, there are abundant examples of specific environmental protection measures—both policies and projects—that show acceptable rates of return even according to narrowly defined benefit-cost criteria. In general, however, the stock of renewable natural resources is rarely considered in a systematic, comprehensive way at the macroeconomic level, where the major strategic planning decisions are made. It is even more rare for explicit linkages to be established between national income accounts and the renewable natural resource base on which so many economies depend.

It is increasingly being recognized that conventional measures of national income, such as gross national product (GNP) per capita, give misleadingly favorable estimates of economic well-being or economic growth (see chapter 3). These measures do not recognize the drawing down of the stock of natural capital (be it renewable or nonrenewable), and instead account for the depletion of resources, that is, the loss of wealth, as net income. Growth built on resource depletion is clearly very different from that obtained from productive efforts and may be quite unsustainable. Unless net capital formation is larger than natural resource depreciation, the economy's total assets decline as resources are extracted or degraded: this appears to be exactly what is happening in many of the poorer natural resource-based economies.

By definition, although exploitation of nonrenewable resources such as oil or coal involves depletion, exploitation of renewable resources such as land and water does not necessarily do so. The complexity of the physical linkages between activities and uncertainty as to the ability of land and water resources to regenerate have in the past tended to mask what has been happening. Costs of natural resource depletion have not been estimated and, along with all other forms of depletion, have certainly not been reflected in national income accounts. Policymakers, relying on GNP as a criterion for national well-being and perhaps being overly preoccupied with short-term considerations may therefore have been lulled into a false sense of security. Indeed, the numbers may be large: for example, rough estimates show that the economic costs of unsustainable forest depletion in major tropical hardwood exporting countries range from 4 to 6 percent of GNP, offsetting any economic growth that may otherwise have been achieved (Pearce 1986:29–31).

The point of the foregoing is not to suggest reform of national income accounts (although they have a number of serious shortcomings in addition to the one noted here). Rather, it is to emphasize the importance of natural resource depletion in the context of overall country planning. That is to say, the macroeconomic impact of natural resource utilization and depletion calls for macroeconomic policies to regulate that use. A critical step in this process is to refine our understanding of relationships between physical events and their economic consequences. Attempts need to be made to quantify the extent of the degradation and to express it in monetary terms where feasible. This would help to highlight the consequences of different patterns of resource utilization for future economic growth and provide a better basis for making strategic decisions concerning the conservation, augmentation, or further exploitation of resources.

The Marginal Opportunity Cost of Resource Depletion

Marginal opportunity cost (MOC) is a useful tool for conceptualizing and measuring the physical effects of resource depletion in economic terms. MOC refers to the cost borne by society of depleting a natural resource, and ideally would equal the price that users pay for resource-depleting activities. A price less than MOC stimulates overutilization; a price greater than MOC stifles justifiable consumption. MOC consists of three elements (see chapter 4): (1) the direct cost to the user of depleting the resource; (2) the net benefits forgone by those who might have used the resource in the future (applicable to renewable resources if they are not harvested on a sustainable basis); and (3) the costs imposed on others, either now or in the future (so-called external costs).

Although the concept has been extensively employed in analyzing the costs of depletable commercial energy resources, calculation of MOC is never easy. This is particularly true of the kind of natural resources discussed here. Nevertheless, several useful efforts have been made, including estimates of the value of agricultural output lost because of deforestation or soil erosion (see chapter 10), and of electricity output forgone because of dam sedimentation (Southgate 1986). Of greater importance than the final result, however, is the discipline required to evaluate painstakingly all physical interrelationships and analyze each in terms of the effects of resource use. In particular it is important to make explicit the tradeoffs or value judgments regarding impacts that cannot be evaluated in monetary terms. These include decisions about the effect on income distribution of alternative patterns of resource use, the impact on vulnerable groups such as indigenous peoples, the preservation of antiquities, irreversible effects, genetic diversity, and the welfare of future generations. The proportion of MOC that may be estimated unambiguously in monetary terms will sometimes be large, sometimes small; the analytical framework for arriving at the point of decisionmaking, however, appears to have universal applicability. MOC may be used effectively as a benchmark to help make judgments about such things as the merits of conservation or protective measures, including investments, regulations, and laws; and taxes, subsidies, and regulated prices of natural resources or their complements or substitutes.

Discount Rates, Irreversible Effects, and Future Generations

Economic analysis is critically important in determining appropriate investments and policies in the environmental area if, and only if, its limitations are recognized. In highlighting the consequences of certain events that cannot be measured in monetary units, and making those consequences explicit in their own narrowly defined terms, economic analysis may be an indispensable aid to good decisionmaking, but it may fail as a discipline if it is pushed too far.

This is well illustrated in the treatment of equity considerations, especially the welfare of future generations. It is often claimed that traditional benefit-cost analysis fails in that discount rates used are so high, that inadequate weight is given to the costs of resource depletion or the benefits of conservation measures to future generations. In fact, manipulation of discount rates is not the answer, for it is inconceivable that one could arrive at a discount rate that satisfactorily reflects the various value judgments and technical parameters involved; for example, private and social time preferences, welfare of future generations, productivity of capital, and economic growth and savings rates. It is much too blunt an instrument for that.

In some circumstances intertemporal choices can be made quite satisfactorily by the use of discount rates that reflect the returns to capital in alternative uses based on fairly short-run market criteria—that is, if there is no reason to expect one generation to be very much worse or better off than another or if irreversible effects are not involved. Gains resulting from projects or activities that pass standard economic tests could, if future societies so choose, be reinvested for the benefit of generations still further in the future. In those circumstances, MOC alone, using market-based discount rates to estimate future costs and benefits, may be used as an adequate benchmark to evaluate investments or policies. If, however, there are irreversible effects (not an unambiguous term, but one that certainly includes elimination of species and any loss of human life, and probably includes desertification) or if future societies are expected to be significantly richer or poorer than the present one, MOC must be supplemented by analysis—possibly quantitative, certainly rigorous—of likely physical and income distributional consequences. The massive uncertainties involved in making predictions about events that will take place many years from now should not deter us from taking such analysis at least as seriously as we now take conventional benefit-cost analysis.

The future generations issue illustrates the role and limitations of economic analysis in natural resource management. It should not, however, be allowed to obscure more immediate environmental concerns or to become an obstacle to their resolution. The urgent problems of the Sahel, for example, suggest that priority be given to resource management and lateral externalities, which have an immediate impact. Irreversible effects are already occurring, and the welfare of future generations will depend to a great extent on measures that improve the well-being of those now living.

The Causes of Natural Resource Degradation

The MOC calculation requires a systematic effort to trace the often highly complex relationships among resource-using activities. The underlying causes of environmental degradation may often be related to activities that at first sight are only remotely connected to the observed effects. If project and policy measures are to be viable, they should be based on a sound understanding of not only the physical linkages among events, but also the equally complex economic, financial, social, and institutional linkages that parallel them. Much work needs to be done in this area: in spite of the massive literature on natural resource degradation, relatively little attention has been given specifically to those points at which institutional or individual behavior plays a key role and at which policy interventions might be feasible.

Further research is needed on the extent of interdependence between man-made activities and natural resource systems; for example, between deforestation, land clearance, and overgrazing on the one hand, and soil degradation and erosion, watershed destruction, and sedimentation on the other. Efforts should be made to quantify the impacts at each stage of the interlinked ecological and economic system in physical or monetary terms in order to determine the points in the system at which it would be most socially profitable to intervene with explicit policy measures. One of the essential elements of this exercise would be to separate out natural events, which, when compounded by human activities, may dwarf environmental damage caused by human activities alone.

Improvement of Physical Data

In the past, establishment of the link between economic—particularly macroeconomic—analysis and environmental considerations has been frustrated by the inadequacy of physical data. This situation is changing rapidly: recent developments in geographic information systems allow not only better assessments of current natural resource endowments and trends in their use, but also better projections of future endowments under various scenarios of economic growth and sectoral output. In particular, remote sensing from space may offer a broad synoptic view, repetitive coverage, and uniformity with respect to the way information is collected. Combined with traditional methods of collecting physical data and the integration of such information with socioeconomic data on population, land tenure systems, and so forth, these developments suggest that systematic linking of macroeconomic and resource planning can indeed become a reality. Economic planners, therefore, must ensure that the collection and analysis of technical information are well focused and geared to operational or policy requirements.

Improved Understanding of Behavioral Factors

We also need to develop a better understanding of individual and institutional behavior as it relates to resource use. This requires a multidisciplinary approach and the analysis of causes that are even more fundamental to the way society works than those already discussed. The range of factors that potentially affect environment-related behavior is awesome: taxes, prices, and subsidies relating to agricultural and forest products, their substitutes, and complements; exchange rate policy; income distribution, land tenure, and property rights; the government and private sector institutional structure; population pressure; educational levels in general and those of women in particular; and the political power structure. All have a role in determining the rate of environmental decay (Pearce 1986; Repetto 1986a). The determination of policies therefore calls for the involvement of the economist, political scientist, sociologist, and anthropologist as well as for legal and institutional expertise, in addition to knowledge of the physical sciences.

Policy Interventions

As noted, the traditional approach to environmental problems is for public authorities to engage in projects such as reforestation and pollution control, which remedy past abuse of the environment, or to prevent degradation by building ameliorative components into industrial projects or irrigation schemes. The technical, economic, and social data and the value judgments needed to make sensible decisions about such investments are also required in designing policy interventions. The empirical and conceptual problems encountered in determining appropriate economic incentives parallel very closely those related to the conduct of benefit-cost studies, and estimation of MOC is equally important in both project and policy analysis. In one area, however, the design of policy interventions is more complex than project analysis: even more than in the case of projects, the success of policy interventions depends heavily on behavioral issues and on the prospects for changing behavior.

Much work needs to be done in this area, but there is already considerable evidence as a basis for concrete recommendations about policy interventions. Examples abound of instances in which government policies, typically in the form of direct subsidies to environmentally harmful activities, are unwarranted even in narrowly defined, traditional economic terms. This is particularly true with regard to forestry policies and to a variety of subsidies used in developing countries to encourage agricultural and forestry activities (Repetto 1985a, 1986b).

Agricultural Pricing Policies

Governments throughout the world intervene in agricultural markets, primarily to keep domestic food prices low. Both direct and indirect interventions are used, most of which tend to reduce agricultural incomes and the ability to invest in conservation measures. The very poorest farmers may be hurt the most: extremely low incomes and the urgency of short-term needs, which imply high discount rates, make the investments required for sustainable output particularly difficult to achieve. Although government intervention may be required because of externalities or adverse effects on income distribution, in general freeing up agricultural markets prices to approximate international levels, tends to be consistent with environmental objectives as well as with traditional, relatively narrowly defined economic goals.

Elimination of Subsidies

Governments, in part to compensate farmers for keeping output prices artificially low, frequently offer a variety of subsidy programs. Many of these are harmful to the environment and are incorrect even in standard economic terms. For example, subsidies have encouraged the excessive use of pesticides, which has not only increased the exposure of individuals to toxic substances, but also led to more resistant strains of mosquitoes and to a resurgence of malaria in many parts of the world. Resistance of other insects to pesticides is also growing and net economic losses—even in the short term—may be a direct consequence of subsidies (Repetto 1985a). Returns have frequently been higher when integrated pest management practices are used—that is, minimal applications of manufactured pesticides, combined with more resistant crop varieties and natural pesticides. Similarly, subsidies for livestock production in the form of credit at preferential rates, tax holidays, and land concessions have often resulted in production that is not justified in either economic or environmental terms (Repetto 1986a and chapter 6).

Governments have also frequently established inappropriate forest revenue systems in which concessionaires do not cover the costs of replacing the exploited forest resources. A variety of subsidies and tax concessions are used, including free provision of infrastructure, such as roads and port facilities; reduced or waived export taxes on processed wood; subsidized credit and export financing; tax holidays and unlimited loss carry-over provisions; and concessional leases (Browder 1985; Repetto 1986a). These subsidies encourage overexploitation, the problems of which are compounded by short-term leases (sometimes for as little as one year) that encourage concessionaires to exploit forests without concern for future productivity. In a number of countries, property rights are automatically conferred upon those who clear the land and use it, thus providing further incentive not to leave the forests untouched.

There are many more examples of incentives that lead to natural resource degradation. Typical means of providing irrigation water, for instance, tend to encourage wasteful use. The introduction of user charges that cover full economic costs, rather than simply operating costs, may do much to improve the situation (Schramm and Gonzalez V. 1977). Perhaps it is too readily accepted that charging for irrigation water on the basis of use presents insuperable administrative difficulties. Pricing of electric power is another interesting illustration: typically, governments require consumers of electricity to pay charges that cover the financial costs incurred by the utility. In this case, however, economic and social costs are often underestimated. Financial costs would be lower than MOC, for example, to the extent that future exploitation of resources costs more than previous schemes (typically true of hydroelectric systems) or that the pollution costs are not fully borne by the utility. Subsidy may be said to exist if the price paid for publicly provided goods or services is less than MOC; it will often be the case that increasing prices beyond what is required to meet the financial objectives of power utilities will improve the efficiency of resource utilization and do so in a way that enhances environmental objectives.

The foregoing examples have a common characteristic: in all cases government policies have adverse effects in both environmental and standard economic terms and offer fairly direct incentives for wasteful environmental management. Greater reliance on natural market forces and removal of the distorting influence of government interventions will in these instances be the appropriate policy stance. The appropriate policy prescriptions in such cases are fairly easy to determine, although vested interests typically make implementation more difficult.

Externalities, the Commons Problem, and Natural Events

Quite appropriately, policy reform in the agriculture and natural resource sectors in recent years has stressed greater reliance on market forces to provide correct signals to producers and consumers. The free market is a good servant but a bad master; since environmental problems often cannot be resolved in an efficient or equitable manner by unregulated market mechanisms, there is no alternative but some form of public intervention. Indeed, the subsidies referred to in the preceding paragraphs might justifiably be replaced by taxes.

Central to natural resource management is, of course, the presence of external effects; thus it is frequently in the private interest of individuals to act in such a way that costs are imposed upon others, who are in no position to demand compensation. Examples are legion: the complex physical linkages among resource-using activities referred to earlier imply a series of external effects that can be controlled only by government intervention.

Interference with market processes is currently a somewhat unfashionable cause to advocate, but many situations call for it. In the case of common land, for example, exploitation of a resource such as grazing land may continue to appear profitable for additional users while actually being disastrous for all; this problem frequently warrants public intervention. Common ownership does not necessarily imply inefficiency: tribal ownership of property in regions with stable populations is often characterized by sustainable farming methods (Pearce 1986). The most serious problems tend to be associated with the use of land or other resources for which ownership is not clearly defined. Measures designed to induce prudent management of such communal resources may include physical restrictions, pricing policies, or the introduction of a variety of property rights, land tenure, and leasing arrangements. The financial and technical assistance and water rights protection given to private pastoral associations in some Western African countries are examples of public interventions aimed at the problem of the commons.

Some incentives to deal with externalities may involve extremely indirect methods. For example, a tax levied on livestock production might reduce overgrazing and land clearance, thereby stemming the rate of soil erosion and and benefiting agricultural productivity many miles away. Ideally, the tax should be such that the livestock producer faces total costs equal to the MOC of his activity, which is determined by, among other things, the effects on soil erosion and on agricultural output elsewhere in the system. Export taxes on logs, taxes or subsidies that vary by crop according to the soil-conserving characteristics of the crop, and subsidization of energy-efficient wood stoves or of kerosene are further examples of interventions that call for carefully weighing the costs and possible adverse effects against the economic and environmental benefits that might result.

Finally, public intervention may be required to manage or ameliorate both the catastrophic and more gradual effects of natural degradation. Measures should be designed in light of the costs and benefits (broadly defined) of the ameliorative action. Damages from natural forces and human activity need to be disentangled, and the set of incentives or other policies designed accordingly. For example, flooding caused by natural erosion and sedimentation might be mitigated by incentives to induce industry or residences to move to less damage-prone areas; to the extent that commercial logging is responsible, the focus should be on incentives designed to improve management of forest resources.

The Administrative Costs of Incentive Systems

A basic argument in favor of relying on incentive systems, is that dealing with widespread environmental degradation on a case-by-case basis, using a benefit-cost approach at the conceptual level and regulation or policing at the practical level, is likely to entail exessive administrative costs. But incentive systems are not costless because to a greater or lesser degree they involve monitoring, policing, and regulation. A system of stumpage fees, for example, may require extensive monitoring; irrigation water charges may need metering. The bureaucratic and legal costs of administering land reform schemes may be overwhelming. The decision as to whether or not a system of incentives (or, indeed, regulations) is worthwhile may be assisted by (and subject to the usual limitations of) a benefit-cost approach. The cost of the incentive system itself—for example, the cost of measuring water consumption and collecting fees from water users—should be compared with the estimated benefits, that is, the savings from the change in resource use resulting from the introduction of the incentive scheme. The magnitude of the savings would depend on the reaction of the users to the price change (price elasticity of demand) and the MOC of the activity to which the incentive scheme is formally applied.

The Need for Parallel Actions

Distribution of Income and Wealth

The interventions discussed in the previous section, some quite direct, some less so, could all conceivably be introduced within existing social and institutional systems. Other changes that affect more fundamental characteristics of the societies concerned are likely to be much more difficult to make. The great inequality in income and wealth in many developing countries, for example, is often reflected in an extremely skewed distribution of land, which by itself may be an obstacle to sound natural resource management.

As population pressure grows, the poor tend to be pushed onto ecologically sensitive areas with low agricultural potential (for example, semi-arid savannas, erosion-prone hillsides, and tropical forests). The situation is aggravated where large farmers respond to growing pressures to expand primary commodity exports and thus enlarge the areas on which cash crops are grown. There is evidence that the large landowners—particularly those engaging in monoculture—do not protect the quality of their land and soil as much as do small farmers who own their land (Pearce 1986:48–56). Although the evidence is not conclusive, intuition suggests that security of land tenure exerts a positive influence on conservation (Feder 1988; Pearce 1986:46). Having inadequate control over the land they farm and little political weight, the poor cannot easily obtain the capital and external information and technology with which they could reverse their plight.

Although land reform is central to the management of natural resources, it is also one of the most difficult issues to deal with. In developing countries the relevant decisions are frequently made by a small, politically influential group with interests in commercial logging, ranching, plantation cropping, and large-scale irrigated farming operations. The prevailing systems of investment incentives, tax provisions, credit and land concessions, and agricultural pricing policies therefore tend to favor those in power. The results are losses for the economy as a whole and damage to the environment and natural resource base.

Institutional Structures

Another obstacle to improvement in natural resource management may be that the institutional structure of the government causes the activities of public agencies to impinge on parties whose welfare is of no concern to them. For example, the costs of a hydroelectric scheme to farmers or indigenous peoples may not be adequately taken into account by a power utility; flooding downstream caused by a river development scheme may not concern a provincial government if the damage occurs outside its borders. Coordination and control of natural resource use in order to mitigate its external effects—in particular to impose incentives that affect several sectors—may require the creation of agencies with wide-ranging authority over certain aspects of the operations of functional ministries in a particular region. It also requires incentives to induce public servants to act in accordance with the common good as well as with the goals of their own agencies. Such structured changes are obviously difficult, but they represent one of the most important public sector management issues facing developing countries today.

Population

It is generally accepted that population pressure is one of the root causes of poverty and natural resource degradation. Accordingly, the success of economic and other incentives will depend in large part on the success of family planning and other population-related policies. For the purposes of the present discussion, there is little to add. Virtually all governments recognize the problem of population growth and are addressing this fundamental issue on both the supply side—by providing family planning facilities—and the demand side—primarily through education. Economic considerations, whether we like it or not, play some role in individual decisions about family size and spacing; the role of governments in influencing those choices by economic incentives or other means is a highly controversial subject, which goes well beyond natural resource issues.

The Role of Women

One population issue of direct concern is the role of women as household and small farm decisionmakers. In many developing societies, women carry the major burden of supporting the household and in performing agricultural work (Noronha and Lethem 1981; Pearce 1986). Without their involvement, natural resource-related policies are unlikely to succeed. In Africa, about 80 percent of subsistence agriculture is carried out by women; men increasingly attend to cash crops or migrate to urban areas. Women normally do not have title to land or adequate access to credit. They may therefore be in no position to take the steps necessary to protect the quality of the land and water resources under their control. The fact that women generally also have less education than men compounds the problem. Equality of educational opportunity, of landownership, and of access to credit are required if decisionmakers at the household and small-farm level are to respond effectively to incentive systems.

Conclusion

An agenda for action is emerging; economics has a major role to play in bringing together and mutually reinforcing environment and development. On the basis of broad assessments of natural resources, economic tools may be employed to help determine the desirability of environment-related projects, their design, and location. Economic analysis is then vital in pinpointing the need for introducing new incentives or removing misguided ones. Used properly, economics can identify the policy instruments necessary for sustainable development.

At the same time, the broader policy focus requires that the traditional economic approach be reassessed and that methodologies be improved and refined. In fact, much could be gained if the tools and concepts already offered by economic theory were to be applied systematically and correctly. Economic analysis must avoid taking a static view and must focus instead on the dynamic nature of the complex environmental and natural resource problems with their multitude of linkages and indirect effects. Many of these effects show up either at distant locations (for example, downstream effects) or in the future (for example, the gradual depletion of soil nutrients), posing a major challenge to economists who must learn first to understand the many coevolutionary processes—the physical interactions plus the human impact—and then to apply suitable economic methods.

Moreover, if economic methods are to be successful, it is crucial that their limitations be understood and continually kept in mind. In particular, it should be recognized that value judgments about distributional and irreversible effects are unavoidable, but quantification in monetary terms of as many variables as possible will crystallize those issues and implicit value judgments that may otherwise be ignored.

Economic assessments and projections will necessarily be fraught with massive uncertainty, given the complexity of the various physical and behavioral linkages in natural resource management. It is therefore imperative that economists recognize the technical limits of their profession and collaborate actively with many other specialists, including engineers, agriculturalists, natural and social scientists, lawyers, and management experts. Above all, a multisectoral and multidisciplinary approach is called for.

Each country should develop its own agenda for action in the following areas.

- Assisted by new technologies, the existing natural resource base should be assessed, trends and patterns in resource utilization identified, and prospects for the future estimated under various scenarios of economic growth by major sector.
- To highlight the magnitude of the problem, the impact of resource depletion on net national product should be estimated.
- Within the MOC framework, the economic and social consequences of major categories of resource use should be estimated.
- The above information should be used in country planning to make judgments about the merits of resource conservation, augmentation, or further exploitation.
- In light of the foregoing, investment programs and areas that require interventions with broad impacts should be identified.
- Government policies that clearly have adverse effects, not only in narrow economic terms, but also in terms of their direct environmental impact, should be eliminated.
- More complex interventions, calling for incentives—price, tax, and subsidy policies—that have an important but often indirect impact on resource use, and that address externalities and the commons problem, need to be designed and introduced.
- Continued efforts should be made to address major underlying causes, not only of natural resource degradation, but of development problems generally, including income and land distribution, population growth, education, the status of women, and institutional reform.

The logic underlying the above agenda is applicable to the handling of environmental problems in general and is clearly consistent with some of the arguments used to justify the principle that "the polluter-pays" (Kneese and Schultze 1975). The experience of developed countries in trying to come to grips with industrial pollution provides no grounds for complacency about the task that lies ahead: indeed, a massive effort—analytical, empirical, and persuasive—will be needed to implement the agenda. There are, however, feasible ways to integrate natural resource issues into economic planning at the national level and therefore to give equal time to environmental concerns. The case for certain types of intervention is fairly straightforward; their impacts will be direct, easily justifiable in conventional economic terms, and environmentally beneficial. Given progress in these areas, there are grounds for optimism that more ambitious steps— more complex, indirect interventions and even policies that address the underlying causes of natural resource degradation—will also be successful.

References

Anderson, Dennis. 1987. *The Economics of Afforestation: A Case Study in Africa.* Baltimore, Md.: Johns Hopkins University Press.

Browder, John. 1985. "Subsidies, Deforestation, and the Forest Sector in the Brazilian Amazon." Report prepared for the World Resources Institute, Washington, D.C.

Feder, Gershon, and others. 1988. *Land Policies and Farm Productivity in Thailand.* Baltimore, Md.: Johns Hopkins University Press.

Kneese, Allen V., and C. L. Schultze. 1975. *Pollution, Prices and Public Policy.* Washington, D.C.: Resources for the Future and Brookings Institution.

Noronha, Raymond, and Francis J. Lethem. 1981. *Traditional Land Tenure and Land Use Systems in the Design of Agricultural Projects.* World Bank Staff Working Paper 561. Washington, D.C.

Pearce, David. 1986. "The Economics of Natural Resource Management." Projects Policy Department, World Bank, Washington, D. C.

Repetto, Robert. 1985a. *Creating Incentives for Sustainable Forest Development.* Washington, D.C.: World Resources Institute.

———. 1985b. *Paying the Price: Pesticide Subsidies in Developing Countries.* Washington, D.C.: World Resources Institute.

———. 1986a. *Economic Policy Reform for Natural Resource Conservation.* Washington, D.C.: World Resources Institute.

———. 1986b. *World Enough and Time.* New Haven, Conn.: Yale University Press.

Schramm, Gunter, and Fernando Gonzales V. 1977. "Pricing Irrigation Water in Mexico: Efficiency, Equity and Revenue Considerations." *Annals of Regional Science* 11(1).

Southgate, Douglas. 1986. "Estimating the Downstream Benefits of Soil Conservation in a Hydroelectric Watershed." Ohio State University, Columbus, Ohio.

World Commission on Environment and Development. 1987. *Our Common Future.* New York: Oxford University Press.

World Resources Institute and International Institute for Environment and Development. 1986. *World Resources 1986.* New York: Basic Books.

3

Environmental and Natural Resource Accounting

Salah El Serafy and Ernst Lutz

Most production and consumption activities have some impact on the physical environment. As economic growth and population expansion have occurred, they have increasingly put pressure on the environment and the natural resource base. Years ago, when the pressure was still light, there may have been some justification for economists to ignore the contribution to economic activities made by the environment, both as a resource base and as a "waste sink," receiving the residues of the production and consumption process. But there is little justification for this now.

Side effects of production and consumption activities have been considered by economists as external effects. But they are external only if a narrow view is taken with no consideration of the impact on the resource system as a whole, which although large, is nevertheless finite and in certain respects subject to great stress. We are now beginning to understand that we have done and are still doing enormous harm to our environment, treating it with disdain as if it were infinite, and that eventually those "external" costs will have to be borne by someone. If a broader view is taken, environmental costs would be internalized within the production processes. In this connection, it is essential to attribute costs and benefits properly, and to distinguish clearly between the true generation of income and the drawing down of capital assets through resource depletion or degradation.

Shortcomings of the Current Measures of National Income

If properly done, income accounting is a crucial tool for economic analysis and policy prescriptions. It can indicate the level of economic activity, its variations from year to year, the size of savings and investments, the limits to what society can consume out of its current receipts, factor productivity, industrial structure, comparative performance, and many other things. Development planners, economists, and politicians thus make frequent use of the national income measure of gross national product (GNP) and variants such as gross domestic product (GDP), net domestic product (NDP), and net national product (NNP) for a variety of purposes. GDP, the most commonly used variant of aggregate income, is essentially a measure of total economic activity for which exchange occurs in monetary terms within a given year. It is valuable mostly for indicating short- to medium-term changes in the level of economic activity, and it is widely used for demand management and stabilization policies. As calculated at present, however, GDP is less useful either for measuring year to year variations in economic activity or for gauging long-term sustainable growth, partly because natural resource depletion and degradation are being ignored. Furthermore, GDP is often used inappropriately as an indicator of economic welfare, frequently without any cautioning about its shortcomings for that purpose. The concept of welfare is much broader than that of a money income measure, and covers many dimensions of subjective well-being other than those involving market transactions and those that can be measured in money terms, particularly for people whose basic material needs have been met.

As most economists know, controversial issues with regard to current practices of national income accounting include the treatment of leisure, household and subsistence production and other nonmarket transactions, and services of long-lived consumer durables. This chapter will not deal with any of those issues; we address only certain environmental and natural resource issues that relate to the proper measurement of income and variations in assets. Because of two key shortcomings in national accounting, GDP, as measured at present, does not adequately represent true, sustainable income. These shortcomings are the treatment of environmental protection costs and the degradation and depletion of natural resources. The fact that these issues are not properly dealt with under the current United Nations System of National Accounts (SNA) represents a serious accounting flaw. As a result, policy advice based on measurements produced under the SNA can be faulty.

The Objective: Measurement of Sustainable Income

True income is sustainable income. This is a key point stressed by Daly (1989) and El Serafy (1981, 1989). True income may be thought of as the maximum amount a recipient can consume in a given period without reducing possible consumption in a future period. This concept encompasses not only current earnings but also changes in the asset position of the income earner: capital gains are a source of income; capital losses reduce income. The essence of the concept of income has been stated by Sir John Hicks (1946:172) as the maximum amount that a person can consume during a certain period and still be as well-off at the end of the period as he or she was at the beginning.

Prudent national economic management thus requires that governments should know the maximum amount that can be consumed by a nation without running down its environmental capital. It is important, therefore, that national income be measured correctly to indicate sustainable income. Adjustments of the SNA appear to be necessary in the two areas noted because they are currently not dealt with satisfactorily: the costs of environmental protection (defensive expenditures) and the depletion and degradation of natural resources.

Defensive Expenditures

Actions are often taken to defend the environment against encroachment by economic activities, and the SNA treats their costs as generating income. Defensive expenditures can be large or small depending on where we draw the boundaries. For the purpose of this chapter we are considering only defensive expenditures against the unwanted side effects of production and consumption (such as pollution), but not those relating to national security, even though similar arguments would apply to the latter. Another possible category would be car repair and medical expenses incurred as a result of traffic accidents. Leipert (1986) has listed yet other costs that might be included, and produced a paper (1987) that measures environmental defensive expenditures for the Federal Republic of Germany.

When expenditures are incurred to redress some or all of the negative consequences of production or consumption, incorporating them in the stream of income generated by economic activity does not make sense. It has been proposed that such outlays should be counted not as final expenditures, as is currently the case, but rather as intermediate expenditures. There are counterarguments against doing this. National defense expenditure, which is one type of defensive expenditure, is much more important in terms of size, but is currently counted as final expenditure. In the case of other production and consumption activities such as those involving tobacco, drugs, and alcohol, one could also justifiably argue for their exclusion from national income aggregates. If environmentally defensive expenditures are defined as intermediate so that they can be deducted from the national income aggregates, national accountants would resist the idea because those definitions would not be consistent with current definitions and conventions under the SNA.

A conceptually different approach looks at resources such as water, air, and soil as natural capital. When such capital is being drawn down or degraded, this should show up as consumption when measuring national income. It should be reflected irrespective of whether or not defensive expenditures are actually incurred either to redress the negative effects or to restore the drawn-down natural capital. The difference between the defensive expenditures actually incurred and the depreciation of the environmental capital would be reflected in the net domestic product. This approach has been proposed by Harrison (1989). A similar conceptual approach has been developed by Peskin (1989), who advocates the introduction of a "nature account" in addition to the standard accounts for households, industry, and government. Aside from the difficulty of reaching a consensus on how natural capital is to be treated conceptually, there is the difficulty of attaching a value to the level of natural capital and environmental services and damages.

Some Aspects of Measuring Pollution within the SNA Framework

Pollution is the discharge of wastes in ways that raise the cost of later activities, harm people, or reduce the enjoyment people get from their surroundings. National accounts can be used to improve environmental policymaking in this important area. Blades (1989) distinguishes four aspects of pollution and considers the extent to which it is possible and useful to measure them within the framework of the national accounts. These are the output of pollutants, the damage of pollution, the costs of abatement, and the benefits derived from such abatement.

Although it may be feasible to use national accounts to measure the output of pollutants, the information so obtained may be too general to be useful for environmental policymaking. With regard to pollution damage, Blades notes that although there are conceptual and practical difficulties in estimating the total costs involved, it would be possible and helpful to identify some of the main costs already included in the national accounts, but not shown separately at present. The costs of pollution abatement are a part of defensive expenditures. They have been measured in several countries and have been incorporated in macroeconomic models in order to show the impact of abatement policies on prices, output, and employment. In this area national accounts would be a valuable tool for environmental policymaking, and Blades considers in detail the conceptual and practical problems of measuring abatement costs. Finally, Blades notes that although it would be interesting to measure the market value of the benefits of pollution abatement, the practical difficulties involved are enormous and it would not generally be feasible to incorporate such data in the national accounts on a regular basis.

Hueting (1989) discusses several ways to deal with defensive expenditures and lists the pros and cons of the various options. He is skeptical about the willingness-to-pay method and advocates that environmental standards be set by considerations of health and sustainable development. The costs of achieving such standards would indicate how far a country has drifted away from sustainable economic development. Although this approach has intuitive appeal, it would be difficult to implement, given the uncertainites about the linkages between production and consumption on the one hand and the environment on the other.

In general, the treatment of defensive expenditures for national accounting assumes greater significance the higher the degree of industrialization of the country concerned. In contrast, depletion and degradation are not directly related to the level of industrialization, but are particularly important for countries that base their economic activities on the exploitation of their natural resources.

Depletion and Degradation of Natural Resources

Under the SNA, there is an evident asymmetry in the treatment of man-made assets and natural resources. Man-made assets—buildings and equipment, for example—are valued as productive assets and are written off against the value of production as they depreciate. Natural resource assets are not so valued or adequately accounted for in most instances, and their loss entails no charge in the national accounts against current income to reflect the decrease in potential future production. A country may be exhausting its renewable or nonrenewable natural resources, and its current income will thus be inflated by the sale of natural assets that will eventually disappear. Differences in recording under the SNA may arise depending on whether a resource is publicly or privately owned. Private companies that take a long view of the natural assets they own often make provisions for the decrease in the capital stock of their natural resources, and in certain countries tax legislation permits the exclusion of such provisions from taxable income. No such exclusion is effected in developing countries where natural resource exploitation is carried out in the public sector.

Underlying this asymmetry is the implicit and inappropriate assumption that natural resources are so abundant that they are costless or have no marginal value. Historically they have been regarded as gifts of nature—a bias that has provided false signals for policymakers. This approach confuses the depletion of valuable resources with the generation of income. Thus it promotes and seems to validate the idea that rapid rates of economic growth can be obtained by exploiting a diminishing resource base. The growth can be illusory and the prosperity it engenders transitory if the apparent gain in income means permanent loss in wealth, that is, if at least part of the receipts is not redirected into new productive investments. As income is inflated, so often also is consumption, and the country concerned becomes complacent about its economic performance. Necessary adjustment to economic policy is then delayed by the seeming prosperity. For natural resource-dependent countries, failure to account for the depletion of the capital stock embodied in natural resources represents a major flaw in the accounting process.

Existing natural capital of geological (nonrenewable) and biological (renewable) resources is needed for agricultural, industrial, and other production. New geological discoveries, as well as recycling and conservation, do not reverse the process of depletion of existing stocks. The newly discovered stocks come from a finite source and they merely extend the time span over which depletion can continue. Depletion of renewable natural resources can have serious indirect effects because a reduction in the stocks or populations of plants and animals may in turn reduce the sustainable flow of resource inputs and ecosystem services. Similarly, crop production at the expense of soil erosion cannot be sustained. Only careful husbandry of environmental capacities can ensure sustainable and potentially larger income flows in the future. The optimistic argument that human ingenuity is bound to find substitutes for the natural resources being depleted may be generally valid, but it would be imprudent for society to base its behavior on such optimism and wrong for economists and accountants not to take rational precautions in case substitution does not occur.

Two main conceptual approaches—the use of depreciation and user cost—have been proposed to deal with the depletion and degradation of natural resources. The principle of depreciation of man-made capital can be applied straightforwardly to the consumption of renewable and nonrenewable resources (Daly 1989; Harrison 1989). Because geological and ecological information on depletion or degradation comes in physical units, these units must be priced or valued in some way before the gross national product can be adjusted to arrive at a corrected net product. Valuation could be based either on the principle of replacement cost where replacement is possible or on the discounted value of the willingness to pay. Present conventions would value the depleted or degraded resources at current prices. If such a correction is effected, based on the full value of depletion, the gross product will remain unadjusted, but the net product will reflect the depreciation of environmental capital that has taken place during the accounting period.

Because the full-value depreciation approach would leave the GDP unadjusted and would eliminate from the net product the entire proceeds from natural resource sales, the user cost approach has been proposed as a way of properly taking into account the depletion of mineral resources. Possession of a natural resource conveys on its owner an income advantage that is denied to those without a natural resource, and a measurement of zero net income, as produced by the full-value depreciation method, is therefore not satisfactory. The user cost approach avoids the difficulties of putting a value on the stock of the resource and relies instead on the conscious assessment of current rates of extraction of the total available stock measured in physical terms. Depending on the rate of depletion and on the discount rate, the gross revenue from the sales of a depletable resource—net of extraction cost—can be split into a capital element or user cost, and a value-added element equal to true income. The capital element represents asset erosion. It has been proposed that capital should be hypothetically (El Serafy 1981) or actually (Ward 1982) reinvested in alternative assets so that it can continue to generate income after the resource had been totally exhausted. Unlike the full-value depreciation method, the user cost approach would alter the reckoning of GDP itself, not just NDP. This method uses current market prices for valuation purposes and is in harmony with accounting principles, but it requires a rule-of-thumb discount rate to convert the capital sales into a sustained income stream. In addition, it is rooted in a proper understanding of the economic meanings of value added and rent, which should not be confused with asset sales. A less satisfactory compromise would combine the two approaches so that GDP would be left inflated with the user cost, whereas depreciation for reckoning NDP would be not at the full value of the resource degradation but at the value of the user cost.

How the net revenue can be split into user costs and true income is explained in El Serafy (1989). One first must decide on a discount rate r, for example, 5 percent. Then one has to determine the number of periods over which the resource is being liquidated. This is simply read from the ratio between total reserves and whatever amount is extracted in the current period. Next, the formula developed by El Serafy (1981) can be used to calculate the ratio of true income X to net receipts (exclusive of extraction costs) R:

$$X/R = 1 - \frac{1}{(1+r)^{n+1}}$$

$R - X$ would be the "depletion factor" (or user cost) that should be set aside and allocated to capital investment and excluded from GDP, while X would represent true, or sustainable, income.

This method is flexible enough to handle changing levels of extraction, movements in the discount rate, and alterations in reserve estimates. Such alterations would include new discoveries, which would change the reserve-to-extraction ratio. In the above formula, alterations in reserve estimates are denoted by n—the life expectancy of the reserve measured in years at the current period's extraction rate. The method is not concerned with the valuation of total reserves, but only with the fraction of the resource being liquidated in the current accounting period, which is valued at current prices. That fraction relies entirely on physical quantities because the price is the same in the nominator and denominator. The method can be adapted to deal with mineral extraction under conditions of deteriorating quality of the product and rising extraction cost, inasmuch as resource owners usually mine the richer deposits first and leave inferior deposits for later extraction, thereby inevitably raising the cost of future extraction.

Like all accounting methods, this method does not indicate an ex ante optimal rate of depletion, but merely mirrors decisions already taken by the resource owner regarding the liquidation of his natural resource. The owner usually determines his extraction rate in the light of many factors, including his expectation of future price changes. If he decides to extract 20 percent of his reserves in one year, then n in the above formulation is equal to 5, and the income content of his net receipts, with a 5 percent discount rate, would be 25 percent and the user cost to be reinvested would be 75 percent. If, however, he extracts only 10 percent of his reserves—that is, over ten years—then he needs to set aside 58 percent of his net receipts for reinvestment, and can enjoy 42 percent as current income. The correction needed to reckon true income out of natural resource sales is higher the closer the resource is to exhaustion and lower the longer its life expectancy at current extraction rates. Were the resource to last 100 years at current extraction rates, only a 1 percent reduction would be necessary in net receipts to arrive at true income with a 5 percent discount rate. The choice of the discount rate materially affects the calculation. A high discount rate, which depresses future against current valuation, raises the ratio of true income in current receipts. But note that alternative investments must be found in which to sink the depletion factor $(R - X)$ so that it can yield that much as a return.[1]

Table 3-1 shows the percentage of net receipts—gross sales minus extraction cost—that represents true income at selected levels of discount rates and years of life expectancy of a natural resource.

Resource Accounting

For resource accounting, data need to be collected on renewable and nonrenewable natural resources, primarily for the purpose of planning long-run exploitation in pursuit of sustainable economic activity. Several industrial countries—among them Canada, France, Japan, Norway, and the United States—have developed resource accounts that are tailored to their specific resource availabilities and policy priorities.

Table 3-1. *True Income (X) as a Percentage of Net Receipts (R)*

Life expectancy of the reserve (years)	Discount rate (r) (percent)		
	2	5	10
2	6	14	25
5	11	25	44
10	20	42	65
20	34	64	86
50	64	92	99
100	86	99	100

Source: Based on El Serafy (1981).

Instead of the expression "resource accounting" the French use the term "patrimonial accounting," which could be described as "accounting for the national environmental heritage" (Theys 1989). The concept is broader than resource accounting because it covers cultural heritage in addition to natural resources. The French resource accounting approach is intended ultimately to relate economic growth to the quantities of natural resources that have to be used up or imported to make economic growth possible. Such a system would also enable a country to optimize the economic value of available natural resources, determine the fraction of GDP that should be set aside for the efficient protection of the environment, and orient economic growth so that it does not threaten ecosystems.

The system when fully developed would be versatile and serve various ends. It could be used to (1) make optimal the use of natural resources as factors of production (for example, through the inversion of a quantitative input-output table that would indicate the intermediate use of natural resources in the productive process); (2) to describe the economic aspects of resource use (for example, by determining which resources would be marketed and in what quantities and values, how to improve the productivity of processing industries to optimize the use of natural resources, and the opportunity costs of alternatives); (3) to treat resources as environmental goods (by taking into account changes in the quality of the environment, costs and benefits of environmental policies, the economic consequences of alternative environmental policies); and (4) to take stock of the national environmental heritage and define the long-term implications of its transformations, so that the environment could be preserved for future generations. Because resources to develop such a system are limited, stress is placed on satisfying the needs of policymakers. Although it would be easier to collect environmental data in the form of flexible reports on the state of the environment and country profiles, the paramount need is to develop a system of environmental accounts so that the information is standardized, exhaustive, summed up in physical and monetary terms, and comparable in time and space. The long-term goal is to match the standards already reached by national (economic) accounting, which make the SNA such a powerful planning tool for economic management.

The French approach is only one among several being pursued in developed countries. (Norway, for example, has been using resource accounts for several years; see Alfsen, Bye, and Lorentsen 1987.) The thrust of the French approach is to build up balance sheets for resources and monitor their change from year to year with emphasis on physical measurements. Such measurements are clearly indispensable; accounting in money terms cannot be undertaken without them. The French approach requires that a comprehensive physical inventory system be in place before any changes can be proposed in national accounting methodology. This is a point of view shared by many, but there are others who want to see national accounting methods adjusted gradually as measurements of parts of the physical environment become feasible.

Linking Environmental and Resource Accounts to the SNA

The SNA does not contain an explicit environmental dimension. The current revisions to SNA were mandated by the U.N. Statistical Commission to simplify and clarify the existing system rather than to propose radical changes in it. This position is being justified by a desire to maintain consistency-in-time series, even if those series contain conceptual shortcomings.

Environmentalists and economists with environmental and resource concerns support several schools of thought about the best approach to the accounting problem. Some advocate environmental accounting in physical terms and have little interest in establishing any linkage with the SNA. Their aim is to use indicators of physical change to influence public opinion and environmental policies. At the other end of the spectrum are those who feel that environmental accounting would not have the same impact unless the accounts were monetized and integrated into the SNA to give an adjusted national income that is more sustainable. We take a middle position: we believe that environmental accounting in physical terms is essential, particularly because the data collected would indicate the speed with which the quantity and quality of natural resources are changing and the direction of change. At the same time, we recognize that monetization, to the extent possible, is also important, and that a linkage with the SNA and an adjustment of the current income measurements are urgently needed. Given the current state of the art, we believe that more work, both conceptual and empirical, is needed before GDP and NDP in the core of the SNA can be replaced by a more sustainable GDP and NDP. That is why, as an interim step, we would encourage the construction of satellite accounts, linked with the SNA, where the adjustments can be made. In other words, the user can compute sustainable GDP and NDP in satellite accounts, which would not threaten the historical continuity of GDP and would have a fair chance of being adopted, particularly to guide policy analysis and prescriptions. If such environmental accounts are adopted, national accountants are bound to take the issues discussed here more seriously and might eventually be willing to adjust the core of the SNA.

Developing Environmental and Resource Accounts for Developing Countries

To ensure that resource concerns are eventually reflected in the SNA and in policymaking, it is necessary to proceed at an operational level so that government officials, national accountants, and economists alike see how, in practice, one can include environmental and resource concerns in the calculations. Certain factors might even facilitate resource accounting work in developing countries, because environmental problems in most developing countries tend to be concentrated and easily perceived. In addition, benefits can be derived from progress made in developed countries and from remote sensing methods of surveying.

Environmental and resource accounting, however, demands a great many data and a great deal of effort, and a plurality of disciplines is required for working on these accounts. The problems are compounded in developing countries by the still limited political demand for this type of activity, because short-run problems are more pressing and the relevant human resources needed are acutely scarce. It is therefore clear that the development of environmental and resource accounts will take time. This fact, however, should not keep statistical and planning officials in developing countries from initiating relevant work now, especially on minerals and forestry, where data are already available.

In the case of Indonesia, Peskin (1989) argues that a local research effort should start right away, supported initially with periodic input from consultants. He also proposes that, ideally, not only environmental but also other important nonmarket factors be considered in an expanded accounting structure.

Repetto and others (1987) have applied resource accounting to fuel and forestry activities in Indonesia. For forestry they estimated the harvesting, deforestation, and degradation net of regrowth; valued the estimates at certain rent factors; and suggested that degradation be treated like depreciation of man-made assets. That is, they proposed reducing the NDP by the estimated amount of depletion. The proposed reductions amounted to over US$3 billion annually for 1979–82, which represents more than 3 percent of the GDP. A similar approach was followed for valuing the depletion of fuel reserves (reservations about this approach and an alternative proposal can be found in El Serafy 1989).

Another empirical study (Magrath and Arens 1987) estimated the costs to the economy of soil erosion in Java. The annual amount estimated was US$350 million to US$415 million, which is somewhat less than 4 percent of the GDP of Java's dryland agriculture. Over 95 percent of these costs are on-site costs of declining productivity. These two studies have made valuable contributions, but it is clear that further empirical work is needed.

A Variety of Approaches but a Common Theme

Some of the papers presented at the Joint United Nations Environment Programme–World Bank series of Workshops on Environmental Accounting over the past few years have put forward incompatible propositions for amending the SNA. This incompatibility should not detract from the central theme (argued in Ahmad, El Serafy, and Lutz 1989) that in their present form the guidelines for income calculation under the SNA leave out extremely important aspects of economic development that should be brought into the accounts. Under present practices the accounts produce readings activity and growth over time that often lead to mistaken policy advice. Such readings frequently exaggerate income, encourage consumption, and promote habits of behavior that cannot be sustained over the longer term.

An interesting argument over what kind of adjustment is desirable and practical is highlighted by the approaches of Harrison (1989) and El Serafy (1989) regarding depletable resources. Both are in fundamental agreement about what constitutes sustainable income and what does not. Harrison would work within the existing framework of the SNA by preserving the definition of final demand used at present, but she would include consumption of natural capital as a parallel entry to consumption of man-made capital with appropriate adjustments to NDP. Further, she argues that income measures should exclude all capital consumption and that net products should therefore be used as indicators of the level of economic activity and its development over time. In contrast, El Serafy would redefine the distinction between intermediate and final demand, maintaining that the sale of natural capital must not be viewed as generating value added and that at least part of that should be excluded from the GDP, as well as from the net product, so that the GDP measurement can continue to be used to describe performance and guide economic policy.

At the other extreme, Norgaard (1989) is altogether skeptical about environmental economists and accountants ever being able to agree on a set of corrections that would simply rectify and fill in gaps in the existing SNA, so that they can end up with one aggregate figure expressed in money terms. He claims that existing SNAs contain contradictions because they are based on conventions and reflect consensus rather than being built on deductive reasoning. Unlike both Daly (1989) and El Serafy (1989), Norgaard does not view sustainability as implicit in the definition of income, but as an "ethical" goal, representing a "separate objective of objective strategies." He sees the concerns of future generations as being undervalued because future generations do not participate in the capital markets of today.

A more conventional view would ascribe such undervaluation not to the absence of future generations from the marketplace, but to the use of too high a discount rate, which has the effect of reducing the value of future net benefits to almost nothing the farther one projects into the future. Because future generations will never be in a position to participate in today's capital markets, the best way to reflect their preferences is to use lower discount rates. The main thrust of Norgaard's argument, however, is against the economic approach to "sustainability of development" based on accounting that relies on market valuations. He espouses "methodological pluralism" in the belief that a multiplicity of perspectives would ensure that "all values are respected" so that decisionmakers have information alerting them to "as many aspects of environmental and resource phenomena as possible" (Norgaard 1989). Norgaard, however, never spells out how such alternative value systems can be established or used, nor does he speculate on the sort of solutions they would bring about.

Various papers in Ahmad, El Serafy, and Lutz (1989) present other areas of disagreement, but the main message of this work is the urgent need to recognize explicitly the shortcomings of current measurements of income and to work toward better measurements and a more sustainable concept of income—common threads in virtually all the contributions.

SNA Expert Group Meeting

In the context of deliberations over whether and in what direction the SNA should be revised, the World Bank presented a discussion note to the SNA Expert Group meeting in Vienna, March 21–30, 1988. The note focused on the two major concerns already elaborated: defensive expenditures and the depletion and degradation of natural resources. With regard to defensive expenditures, the Bank recommended (1) that the experts discuss and decide as a matter of urgency which expenditures are to be included under this category; (2) that they show these expenditures as a separate line item; and (3) that the Group decide where the adjustments are to be made and whether the adjusted aggregates are to be shown in satellite accounts or in the main system of production, consumption, and accumulation accounts.

With regard to depletion and degradation, the Bank (1) urged the adoption of SNA-linked satellite accounts where adjustments for resource depletion and degradation can be made; (2) recommended that the Expert Group warn countries in which a significant portion of GNP derives from the depletion and degradation of natural resources, and that their current income, as calculated at present, over-estimates sustainable income (policymakers should particularly be alerted to this fact); and (3) requested that the Group encourage work on estimating costs and benefits related to the exploitation of depletable natural resources.

The Expert Group acknowledged the importance of the two issues raised and expressed concern about the way the environment was being accounted for under the SNA. Although the experts were not yet ready to advocate changes in the core accounts, they were prepared to support the construction of satellite accounts to be linked to the SNA. Rearranging the core accounts would require a conceptual consensus that was not likely to materialize during the ongoing revision, which has to be finalized by 1991. To attain that consensus, it would be necessary to undertake a functional breakdown in the central framework of such activities as health and research and development in addition to considering the environment. Rearranging the core accounts for just one function would not be justified even if such rearrangement would be beneficial from an analytical as well as an accounting point of view. The experts stressed that valuation issues were very difficult and required further work before standard valuation techniques were agreed upon.

Among the many points made during the discussion were that (1) shortcomings of the GDP as currently measured should be emphasized in the revised SNA manual known as the "Blue Book"; (2) national accountants and environmental economists should tackle the problem jointly; and (3) approximations required for environmental accounting might not be any worse than some conventional estimate methods already sanctioned by the SNA. Such issues and others related to them merit further discussion: a possible venue for this might be the SNA Expert Group meeting scheduled for January 1989, which would deal with inventories and physical assets and their treatment in balance sheets and reconciliation accounts. As a concession to the environmental community the Expert Group chairman stated that if and when adequate experience had been gained with satellite accounts and various conceptual and valuation issues had been resolved, an SNA revision would be called for even if that proved to be much sooner than envisaged.

Ongoing and Projected Work

Although efforts are being exerted to refine concepts and develop approaches to take care of environmental concerns in national income accounting, some of the principles proposed need to be applied to a set of case studies before they can be generally accepted. Environmental concerns are already being integrated into the World Bank's country economic and sector work. Gone are the days when natural resources could be treated as extraneous to economic activity. Resource issues must be integrated in macroeconomic analysis and policies so that resources can be managed efficiently in the service of sustainable development.

The integration of environmental and resource concerns into the SNA, albeit initially via satellite accounts, is an important part of this endeavor. An in-depth review is also planned of the resource accounting done by industrial countries (such as Canada, France, Japan, Norway, and the United States) so that lessons can be drawn from that work. As the case studies are undertaken, perhaps over a two- to three-year period, parallel work should proceed to produce internationally standardized methodologies, perhaps in a handbook on environmental accounting within the series of the SNA. At the March 1988 meeting the Expert Group stated that it would be invaluable as an interim measure if an international organization were to piece together what is known about environmental and resource accounting, with emphasis on problems capable of solution, in order to minimize the duplication of effort and resources that would result from individual countries pursuing the subject independently.

Note

1. J. M. Keynes (1936:ch. 6) first introduced the concept of user cost—in relation to capital equipment—defining it as the "maximum net value which might have been conserved . . . if it [the equipment] had not been used." He described user cost as "one of the links between the present and the future." Project analysis of depletable minerals has also made use of the concept of user cost at the micro level (see, for example, Schramm 1986).

References

Ahmad, Yusuf J., Salah El Serafy, and Ernst Lutz, editors. 1989. *Environmental Accounting and Sustainable Income.* Washington, D.C.: World Bank.

Alfsen, Knut, Torstein Bye, and Lorents Lorentsen. 1987. *Natural Resource Accounting and Analyses: The Norwegian Experience, 1978–1986.* Social and Economic Study 65. Oslo: Central Bureau of Statistics of Norway.

Blades, Derek W. 1989. "Some Aspects of Measuring Pollution within the SNA Framework." In Ahmad, El Serafy, and Lutz (1989).

Daly, Herman. 1989. "Toward a Measure of Sustainable Social Net Product." In Ahmad, El Serafy, and Lutz (1989).

El Serafy, Salah. 1981. "Absorptive Capacity, the Demand for Revenue, and the Supply of Petroleum." *Journal of Energy and Development* 7(1):73–88.

_____. 1989. "The Proper Calculation of Income from Depletable Natural Resources." In Ahmad, El Serafy, and Lutz (1989).

Harrison, Anne. 1989. "A Possible Conceptual Approach to Introducing Natural Capital into the SNA." In Ahmad, El Serafy, and Lutz (1989).

Hicks, John. 1946. *Value and Capital*, 2d ed. Oxford, U.K.: Oxford University Press.

Hueting, Rofie. "Options for Dealing with Defensive Expenditures." In Ahmad, El Serafy, and Lutz (1989).

Keynes, John Maynard. 1936. *The General Theory of Employment, Interest and Money.* London: Macmillan.

Leipert, Christian. 1986. "Social Costs as a Factor of Economic Growth." *Journal of Economic Issues* 2(1):109–31.

_____. 1987. "Defensive Ausgaben in der Bundesrepublik Deutschland 1970 bis 1985: Absolute Werte und Relations zahlen mit dem BSP." Berlin. Processed.

Magrath, William, and P. Arens, 1987. "The Costs of Soil Erosion on Java—A Natural Resource Accounting Approach." World Resources Institute, Washington, D.C. Processed.

Norgaard, Richard. 1989. "Issues Related to the Linkage of Environmental and National Income Accounts." In Ahmad, El Serafy, and Lutz (1989).

Peskin, Henry. 1989. "Environmental and Non-Market Accounting with Some References to Indonesia." In Ahmad, El Serafy, and Lutz (1989).

Repetto, Robert, Michael Wells, Christine Beer, and Fabrizio Rossini. 1987. "Natural Resource Accounting for Indonesia." World Resources Institute, Washington, D.C. Processed.

Schramm, Gunter. 1986. "Practical Approaches for Estimating Resource Depletion Costs." In *Natural Resources Economics and Policy Application,* edited by E. Miles, R. Pealy, and R. Stokes. Seattle: University of Washington Press.

Theys, Jacques. 1989. "Environmental Accounting and Its Use in Development Policy." In Ahmad, El Serafy, and Lutz (1989).

United Nations. 1968. *A System of National Accounts.* U.N. Publication Sales no. E.69.XVII.3. New York.

Ward, Michael. 1982. *Accounting for the Depletion of Natural Resources in the National Accounts of Developing Economics.* Development Centre Publication CD/R(82)3010. Paris: Organisation for Economic Co-operation and Development.

4

Marginal Opportunity Cost as a Planning Concept in Natural Resource Management

David Pearce and Anil Markandya

There is now widespread appreciation of the fact that the economic fortunes of many developing countries are inextricably bound up with the state of their natural environments. In particular, there is serious concern that primary renewable and quasi-renewable resources—soil, water, and forest biomass—are being "mined" and depleted to the point of actual or potential nonrenewability. This concern focuses on (1) exploitation that causes irreversible phenomena such as desertification and eliminates the option to rebuild the resource base to some policy-determined level; (2) the complex linkages among renewable resources, which spread the costs of depletion of one resource throughout the ecological system; (3) the speed at which these effects occur; and (4) the immediate and future cost of human misery, especially among the rural poor, because of resource degradation. Quantified descriptions now abound, as do pleas to take constructive action (see, for example, World Resources Institute 1986; Repetto 1986a, 1986b; Bartelmus 1986; Holdgate 1982; Warford 1987; and Pearce 1987a).

Natural resource degradation (NRD) affects developing countries in two general ways. First, it affects their direct dependence on natural resources. Whereas developed economies have "roundabout" technologies in which the relationships among final products and natural resources are often obscure and complex, in developing countries households and agricultural sectors tend to use natural resources directly and on a daily basis. Examples include the direct reliance on wood fuels, the use of rivers and lakes as sources of water, and even the use of wildlife for food. To illustrate, only 24 percent of the rural sector in Africa had house connections to water supplies in 1983, although this was a marked increase over the 10.8 percent figure for 1970. Water connec-

The authors wish to acknowledge assistance from the World Bank and the U.K. Economic and Social Research Council for earlier work on which this paper is based.

tions to urban households in Africa increased in number over the same period, but as a percentage of the urban population, the figure fell from 63.5 percent to 59.6 percent (World Health Organization 1986). Use of traditional fuels (wood fuels and other biomass) as a percentage of total primary energy supply is above 90 percent in Malawi, Nepal, and Tanzania, and between 70 and 80 percent in Ethiopia, Guinea-Bissau, Niger, Paraguay, and Sudan (authors' calculations from World Bank and United Nations Development Programme Energy Assessments). The very existence of subsistence agriculture underlines the direct dependence on soil fertility, rainfall, and natural or managed irrigation water. The depletion of these renewable resources is likely to have detrimental effects on the populations that rely on them.

Second, the development process will be affected indirectly by NRD. Development is best indicated by a vector whose components include real incomes per head, health, education, and other basic needs. NRD affects a number of these components of development, as where contaminated water spreads diseases. It also affects more traditional indicators of change such as per capita real incomes, especially if these are (properly) construed in sustainable terms. That is, resource depletion may well yield temporary gains in real income in the same way that anyone can borrow from a capital fund, but continued depletion is likely to result in medium- to long-run income losses, depending on the dynamics of the development process. Examples of losses include depreciation in forest stocks due to nonsustainable logging practices (Repetto 1986c); hydroelectricity losses due to dam sedimentation, which arises in part from nonsustainable forest clearance for agriculture and wood fuels (Sfeir-Younis 1986; Pearce 1986a); and crop losses due to the use of natural fertilizers—livestock dung and crop residues—as fuels in the face of wood fuel shortages. In short, the renewable resource base of the economy ceases to be a dispensable input to the development process and must be viewed instead as both a condition of and an integral part of development. Complex issues arise about the optimal size of that renewable resource stock (Pearce and Markandya 1987). It is clear, however, that a number of countries have depleted their critical renewable stocks to such a great extent that their development potential is at risk. This overall message is recent but now familiar (see, for example, International Union for Conservation of Nature and Natural Resources 1980; Brown 1987).

In this chapter we show how the tools of analysis that are already familiar to natural resource economists can be used to explain certain features of the NRD process and point toward policy measures to correct them. The concept we use to investigate these aspects of NRD is the marginal opportunity cost (MOC). In calculating MOC, we seek to measure the true cost of an action or policy that depletes a unit of renewable resource in a way that inhibits the natural regenerative process; that is, the resource is managed in a nonsustainable way. Because the concept is no different from the more familiar one of social cost, it may be argued that it is nothing new. We accept that the concept is familiar—at least to economists—but argue that it is suggestive and thought-organizing in several interesting ways. In particular, we argue that it tells us something about (1) the social pricing of natural resources in developing economies, (2) the spatial dimensions of the appraisal of project, and (3) the intertemporal dimensions of the NRD process. To set the scene further, we briefly consider the nature of the links between economies and their natural environments, and the emerging models of development that focus on the renewable resource base.

Linkages between Economy and Ecology

Since Kenneth Boulding's (1966) seminal spaceship earth essay, there has been general, but not universal, awareness that the linear economy approach of economics textbooks is a misleading abstraction. Linear economies consist of production and consumption sectors, with the process of maximizing the social utility of consumption being constrained only by the rate at which resources can be transformed into production and consumption. Boulding observed that the laws of conservation of mass link the natural resource base, which feeds the production sector, to the emission of wastes into the receiving natural environments. Because the environments have limited, though variable, waste-assimilation capacities, there are constraints on the rate of resource transformation (in addition to those cited in conventional economic models), and on the time period over which such transformations can take place. These constraints are beyond the limitations set by time and social organization. The limits set by the relationship between the emission of waste and the capacity of the enviroment to assimilate that waste is the analogue of rules for the sustainable use of renewable resources, namely that harvest rates should not exceed natural or managed yields. The linear economy is replaced by one in which both the flows of finance and the flows of materials and energy are circular. Moreover, the two layers of the system interact: economic and ecological systems cannot be separated.

Boulding further acknowledged the economic importance of ecological cycles familiar to any life scientist—for example, carbon cycles, hydrologic cycles, and nutrient cycles. His focus was planet earth as a global ecosystem, but others have shown how the principles apply to regional and national economic systems (Kneese and others 1970; Hafkamp 1984). In the context of a developing country figure 4-1 provides a stylized summary of some of the main linkages between the

Figure 4-1. *Linkages between a Developing Economy and Its Ecosystem*

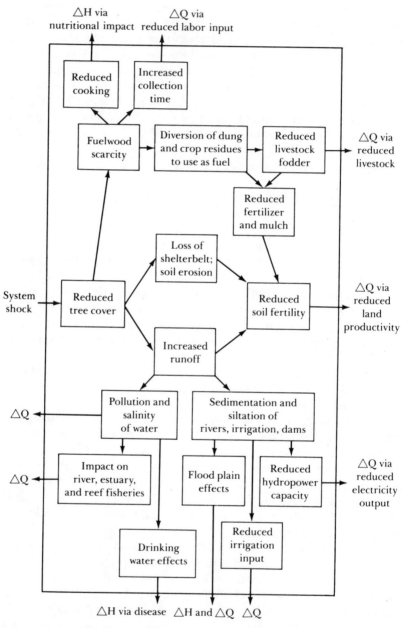

Note: △Q refers to effects that show up in measured or measurable indicators of development. △H refers to effects on health.

economy and the ecosystem as they relate to renewable resources. A system shock, perhaps in the form of agricultural colonization of a hitherto afforested area, is seen to have various cumulative effects. Loss of tree cover increases soil erosion, erosion adds to watercourse sedimentation, which reduces electricity output and raises flood plains, and so on. The direction and scale of these effects depends on a second layer of the interaction between the economy and the ecosystem, namely the level of social organization. Thus the social response to deforestation could be the development of stable agricultural systems that prevent soil erosion.

The essential point is that economic development is not independent of the renewable resource base. In Norgaard's terms (1984, 1987), economy and ecosystem are coevolutionary—development of one requires the harmonious evolution of the other. In the language of neoclassical economics, the system shock generates external effects. But the theory of external effects needs to be broadened to take account of the following:

- The externalities may well be pervasive because of extensive ecosystem linkages accompanied by the phenomenon of direct dependence.
- The externalities may well extend over wide geographic areas, although the watershed appears to set a reasonably bounded system before effects are dissipated.
- The externalities have a temporal aspect in that resource degradation now precludes the benefits of future resource use.
- Development, insofar as it contributes to sustainable growth in social indicators, is often the source of externality, which, in turn, is compounded to feed back negatively on development.

MOC seeks to identify and measure the true social costs of actions and policies such as deforestation. As such, it has to begin by identifying the relevant ecosystem linkages.

Alternative Models

It is fashionable to speak of the social goal in developing economies as sustainable development. Unfortunately, the term has come to mean all things to all people, not least because the term development is itself a value-loaded concept. Four aspects of development are discernible: (1) traditional models of steady-rate growth such as those found in the theory of economic growth (see Jones 1975; Hacche 1984); (2) development paths that are resilient to external shocks such as drought, population change, and exogenous events such as world price changes in commodity and input markets (Conway 1983); (3) development with an explicit bias toward the rural poor and immediate benefits (Chambers 1983); and (4) development that honors certain rules of the game, notably the physical requirements for the sustainable use of resources such as land, water, biomass, and the assimilative capacity of the environment (Pearce 1986b, 1987b; Page 1977).

These alternative views are discussed in detail in Pearce (1988b). The fourth variant, which has its heritage in the work of Boulding, is the most suggestive and has direct relevance to the processes of NRD in developing countries. Moreover, when translated to a dynamic context, it has affinities with the few attempts that have been made to develop an ecological theory of economic change, notably the major neglected work of Wilkinson (1973), which has recently been revived and tested by Common (1988). In Wilkinson's theory, development occurs only as a result of some disequilibrium in the relationship between the economy and its population and the natural resource system supporting it. But the kinds of change that take place in response to disequilibrium offer no guarantee that the economy will get on to a stable path—it might also collapse or enter a depression stage. In the modern context, the risk of failure may be quite high because the social forces responding to disequilibrium are accompanied, and often swamped, by policy responses from government and other authorities. If the two are not in harmony, the risk of failure is high. Up to the ecological boundary set by the endowment of natural resources—especially renewable ones—economic change may be steady. When the boundary is encountered disequilibrium occurs. Pursuit of the traditional development path can easily lead to short-run gains, notably as the renewable resource stock is depleted: borrowing from the natural capital base takes place. In the absence of major technological breakthroughs, the development path is short-lived (a few decades rather than many).

A sustainable development path occurs only if the ecological boundary is shifted. Mechanisms for doing this include the application of relevant technology, management of renewable resources to secure higher natural yields, investment in assimilative capacity, recycling, and a switch to borrowing from exhaustible resources such as oil and coal.[1] Policies to shift the boundaries may be risky—technology may create as many problems as it solves; for example, when agricultural mechanization is inappropriate for the depth of soil. In this sense, development is no more guaranteed than it is in any of the traditional models of economic growth. If the model gives ecological factors a central role in the development process, however, it becomes clear that the risks of development are greatly magnified if these factors are ignored.

How does this sketchy outline of an ecological development process fit with the concept of marginal opportunity cost? Forcing attention on the components of MOC highlights their role in the development process is highlighted. As we shall see, MOC is future oriented and spatially oriented. Ecological boundaries enter the picture through a user-cost component, which measures the benefits forgone in the future by depletion of the resource today. Arguably, even this user-cost component has to be modified to ensure that the full costs of nonsustainable actions today are accounted for (Pearce 1988a), but the general point is adequate for current purposes.

The Concept of Marginal Opportunity Cost

When a small amount of a natural resource is used up, the true value of that resource is measured by the marginal opportunity cost. In this definition, the word "marginal" occurs because the calculation is done for a small change in the rate of usage. Economists frequently use marginal concepts in determining the rules for resource allocation and in measuring scarcity. The reason for this is that the appropriate level of use of a resource can often be calculated by equating the marginal cost of that resource with the marginal benefit derived from its use. If both marginal cost and benefit can be calculated, then we can check to see whether they are equal. If the cost exceeds the benefit at the margin, then this indicates that the resource is overexploited and its use should be cut back. Conversely, if the marginal benefit exceeds the cost, then increasing exploitation slightly should be beneficial.

In addition to working out the appropriate level of use of a particular resource, calculations of marginal cost can also be of use in evaluating public investment projects and government regulations. Such activities often involve small changes in the composition and level of the natural resource base of the country. If the costs of such changes can be measured, they can be included in the overall calculus of costs and benefits, from which a decision on the suitability of the investment or regulation can be made. In this respect, the marginal opportunity cost (described in detail below) is the same as the marginal social cost of any input used in or affected by a project or regulation. The difference is that it refers to the marginal cost of a natural resource and is calculated somewhat differently.

Although we have argued that MOC is generally the correct measure of scarcity, the appropriate concept is not always a marginal one, particularly when the policies being considered involve large changes to the stocks of natural resources. In this case, the value of a small change in the resource, suitably scaled up, will not be an accurate measure, and what is required is a comparison between the value of the total stock before and after the change. In addition, changes in the values of related resources and commodities should be compared before and after a change. Typically, such global comparisons will be required in the context of natural resource management when evaluating the consequences of ecological disasters. In such cases, the notion of a marginal disaster is something of a contradiction in terms. This qualification to the use of marginal measures is important and should always be borne in mind. However, the most relevant and frequently used concept in the management of scarce natural resources is still MOC.

Opportunity cost refers to the best alternative use to which particular resources could be put if they were not being used for the purpose being costed. These costs have three components. First, there is the direct cost of the activity. Extracting natural resources requires labor and materials; for example, cutting down a tree may require one person-day of labor. Suppose that the same person-day could, if devoted to another activity, produce goods and services to the value of $X. Then the opportunity cost of that labor is said to be $X, and that is the figure that should be entered into the direct-cost calculation. The relationship between the opportunity cost as described above and what is actually paid to the worker can be quite complex and involves a number of considerations that are not relevant here. In general, however, actual payments for inputs and commodities will need to be adjusted in the light of taxes and market imperfections in order to obtain their opportunity cost. Such a process is sometimes referred to as shadow pricing.

The second component of MOC is the external cost. As explained earlier, these costs arise because changes in any single component of the natural resource base affect the other components of that base and the efficiency with which other economic activities can be conducted. For example, deforestation may result in soil erosion and river and reservoir siltation. This could affect agricultural output, electrical output, and the quantity and quality of drinking water now and in the future. Such impacts are measured in terms of the value of the activity or commodity in its alternative use. In the above example, reduced agricultural and electrical output and drinking water have a cost equal to the sum of the consumers' willingness to pay for these commodities. The fact that some costs occur in the future means that we discount them, using a discount factor to make them comparable to present day costs. If, for example, the social discount rate is 5 percent a year, then a cost of $1.05 in one year's time is equivalent to a cost of one dollar today.

To determine these external costs, one has to look at the data on the actual prices paid for the commodities concerned and the nature and structure of taxes that apply to them and obtain more general information on the determinants of the demand for those commodities. The determinants of demand are relevant in finding out whether there is excess supply or demand for the items concerned at present, and in ascertaining what the future demand for those items is likely to be. Although information on these issues is imprecise and difficult to obtain, a useful approximation to the value of the marginal external cost can be calculated in many cases.

As stated at the beginning of this chapter, the external costs of particular relevance are those that arise when the resource is being exploited on a nonsustainable basis. Spillover effects that arise in sustainable use are likely to be small and may be internalized. This means that, with repeated use, people will eventually realize that the exploitation of natural resources has an impact, and its costs will appear as a direct cost. External costs also arise from the sustainable use of a resource, but we argue that they are of secondary importance. The difference between sustainable externality and nonsustainable externality is discussed in Pearce (1988a).

The final component of MOC arises from intertemporal considerations. Initially, let us suppose that the resource we are dealing with is not renewable but fixed in supply, so that any positive rate of exploitation will imply eventual exhaustion. In that case, using one unit of the resource now implies that it will be unavailable in the future. This places a scarcity premium on the resource, the amount of which will depend on how large the stock is relative to the rate of exploitation, how strong future demand is relative to the present demand, what substitutes are likely to be available in the future and at what cost, and what the discount factor is. For details of the calculation of this premium, which is defined as the user cost, the reader is referred to Munasinghe and Schramm (1983). Assume, for example, that under current expectations a resource that has a direct plus external cost of $1 per unit will be exhausted in ten year's time. At that stage, it will be replaced by a substitute that has a price of $2. Then, at the moment of exhaustion of the first resource, we would expect it also to have a price of $2. Otherwise, either the substitute would be cheaper, in which case no one would buy the fixed resource, or the substitute would be more expensive, in which case no one would want to buy it.

The present value of $2 in ten year's time will depend on the rate of discount. If the rate is 5 percent a year, then this value is $1.23—that is, $2/(1.05)^{10}$. This means that one unit not used today but used in ten year's time has a value of $1.23. Consequently the opportunity cost of consuming the unit today can be said to be $1.23. Because we have calculated the marginal direct and external costs as one dollar, there remains a premium of 23 cents to be added in order to obtain the overall MOC. This last component, which is the user cost, clearly depends on a large number of factors. The discount rate is obviously a key variable, but so are the future price of the substitute and the time at which it comes into use. Hence, uncertainty about future developments and prices will play a significant part in determining user cost.

The above discussion refers to an exhaustible resource. If the resource is renewable, and if its present and future use is likely to take place on a sustainable basis, then any of the resource that is exploited today will be exactly replaced through natural or managed regeneration. In that case, there will be no scarcity premium to be added to the present direct and external costs. The present situation in many countries, however, is not one of sustainable use. In some countries, complete exhaustion of the resource is the most likely possibility. In that case, we can treat the resource as exhaustible and calculate the user cost as outlined above. In other cases, it might be possible to arrest the rate of exploitation so that at least a minimum stock (that is, one that prevents major ecological disasters) is preserved. Doing so will take some time, but once that minimum stock is attained it would seem logical for the authorities to maintain it. Under this scenario, a scarcity premium can be attached to current usage of the resource because future use is going to be restricted and future prices are going to be higher—other things being equal (see Pearce and Markandya 1987).

To sum up, MOC is made up as follows:

$$MOC = MDC + MEC + MUC$$

where MDC is the marginal direct cost, MEC the marginal external cost, and MUC the marginal user cost. A considerable amount of information is required for each of the components and particularly for the last two. MEC requires details of the engineering and scientific relationship between natural resources and economic activities. It occurs mainly when the resource is being exploited on a nonsustainable basis. MUC requires expectations to be formed about future patterns of exploitation and about future developments in the demand for natural resources and the supply of substitutes for these resources. For renewable resources MUC arises only when the resource is being used on a nonsustainable basis.

The Uses of Marginal Opportunity Cost

We have defined MOC and attempted to show how it relates to the wider theory of the relationship between an economy and its ecosystems, all in the context of the development process. We now illustrate the uses to which the concept can be put.

MOC as an Organizing Concept

Because MOC is a logical extension of the economist's traditional preoccupation with marginal cost pricing to achieve allocative efficiency both intra- and intertemporally, it serves as a mechanism for listing the kinds of costs and benefits that need to be considered when evaluating investments to counteract NRD. Similarly, NRD is frequently a side effect of investment policy. In the investment context, MOC amounts to a marginal version of cost-benefit analysis in general. For example, consider an investment to counteract desertification. A package of measures is introduced that includes shelterbelt forestry, soil management techniques, and tree growing for livestock fodder and fuel. The benefits will show up as savings in time spent collecting scarce fuelwood, increases in farm productivity arising from improved livestock and improved soils, and the avoidance of desertification. These benefits are the mirror images of the component cost items in MOC.[2] Similarly, the true costs of allowing NRD to continue are indicated by MOC, just as they were at the macroeconomic level when considering the costs of NRD in terms of conventional development objectives.

MOC and Shadow Pricing

MOC also has implications for shadow pricing, that is, the prices that reflect the true state of scarcity of the natural resources in question. In the context of developing countries, the practice of shadow pricing for cost-benefit analysis tends to be based on the methodologies developed by Little and Mirrlees (1974) and Squire and van der Tak (1975). In broad terms, this requires that inputs and outputs be valued according to their opportunity costs. For goods that either are or could be internationally traded, the relevant shadow price is the border price, the price that could be obtained by exporting a good or the price that has to be paid if importing it. If a ton of oil is consumed domestically, for example, what is forgone is the foreign exchange that could have been earned if it were exported. For goods that are not traded internationally, the marginal cost of supply is the relevant shadow price because this reflects the cost of the resources used up in that supply.

MOC now replaces the usual marginal cost concept and thus becomes the shadow price for nontraded goods and inputs. For tradable goods, the border price remains the correct shadow price. Figure 4-2 shows the relationship between MOC and border prices. If the border price exceeds MOC, then the tradable goods sector should be expanded, because the marginal benefit of that expansion (the border price) exceeds the true marginal cost of the expansion (the MOC). A frequent complaint in developing economies is that the world market does not adequately compensate for the true costs of supply; the exporting nation bears all the costs of NRD and these outweigh the foreign exchange revenue. In figure 4-2 this implies that the nation is operating to the right of Q^*, the point at which marginal benefits equal marginal costs of supply. If so, the tradable goods sector is too large and should be contracted.

Figure 4-2. *The Relation between Marginal Opportunity Cost and Border Prices*

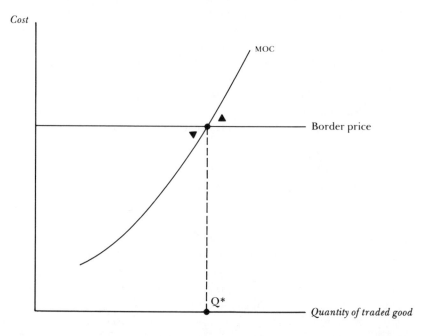

MOC as a pricing principle again forces attention onto the externalities associated with NRD. Moreover, it guides actual pricing policy in providing incentives for allocative efficiency. It is widely argued that pricing policy should first be concerned with the elimination or reduction of subsidies, which encourage excessive resource use. As an allocative principle this is correct, that is, proper marginal cost pricing will tend to have incidental environmental benefits if it reduces improper and wasteful uses of resources. Failure to set water charges for irrigation on the basis of user benefits, for example, is a notorious cause of inefficiency in the agricultural sector (Repetto 1986a). But it may well be that the divergence between marginal (private) cost and MOC is still very large even when marginal cost pricing is approximated.

MOC and the Unit of Account

The externality component of MOC reflects the ecological linkages among sectors. Figure 4-1, shows that the effects can show up in locations quite distant from the initial act of NRD. Project evaluation thus needs to consider effects within a spatial unit of account—the watershed. The watershed as a management unit is well established in both developed and developing economies. Less practiced is the analysis of the effects of specific investments on an entire watershed. An example of neglected effects from the construction of hydropower reservoirs is the magnet effect of the reservoir on human settlement. Reservoir construction attracts settlement, which in turn may lead to land clearance for agriculture. Agriculture causes soil erosion, which then contributes to siltation rates in the reservoir. By itself, MOC does not detect these effects, but it will when allied to an understanding of the watershed's ecosystem.

Discount Rates

Many of the external effects from NRD will show up in the future, perhaps to be borne by those involved in the process, perhaps by future generations. The user cost component will tend to be borne by future generations. As noted previously, the nonsustainable harvesting of renewable resources will have future costs, in terms of both externalities (MEC) and forgone benefits (MUC). In both cases the costs of future losses are expressed in terms of present value, that is, how they are viewed now. The size of these components will therefore be partially determined by the discount rates, which may reflect the high rates of return in underdeveloped agricultural sectors or the high interest rates in credit markets. Yet in both cases it is the process of NRD that actually contributes to the high discount rates. NRD will make risk premiums very high in actual credit markets, and if NRD is severe, it will generate high time preference rates as the search for immediate gains to prevent starvation becomes all the more urgent. Moreover, if high discount rates are adopted there is a paradox, for they will reduce the MEC and MUC components of MOC, making it seem that NRD is less serious. The optimal level of investment in avoiding NRD thus appears less. NRD both creates high discount rates and is made worse by those discount rates. The fallacy lies in using market rates or inferred rates of time preference to guide the choice of social discount rate. High discount rates are, in many respects, a restatement of the NRD problem.

MOC and the National Accounts

The national accounts of a country attempt to measure the value of the goods and services produced in that country and to show the division of that value among various categories of households. This exercise is of course extremely difficult to do accurately, especially in developing countries, and items may be incorrectly included or omitted from the final calculation of national income. Natural resources that are marketed and those that are used outside the cash economy need to be considered separately. The former will usually be priced in terms of their direct cost alone (MDC), and the value included in the national accounts will reflect the equality between MDC and the consumer's willingness to pay for the resource used. The true value, however, is less than that. This is because consumers of the resource impose costs on other agents in the form of restricted output or higher prices now and in the future. We measure these costs as MEC and MUC. To obtain the true value to society of the present consumption of the natural resource, we should subtract from the value of the marketed output MEC plus MUC times the number of units consumed. MUC acts here as an analogue to the depreciation factor on capital and therefore should be subtracted not from gross income but from net income. MEC is a mixture of present and future costs, and its treatment with respect to gross and net income is not clear unless each case is looked at in detail.

In some cases, government expenditure might be undertaken to mitigate some of the external effects of NRD. Such expenditures are frequently included as final consumption by the government, and appear in the national accounts as such. This, of course, is incorrect, because they are a cost of consumption of natural resources and should be properly treated as an intermediate input and netted out of the national accounts. These expenditures are easier to identify once we start to measure MEC.

When natural resources do not go through the cash economy, they are either excluded from the national accounts or included on an estimated basis. The most likely situation is that users of the resource bear the marginal direct cost and equate that cost to their marginal willingness to pay for the resource. Hence that cost is the direct value of the use of the resource. However, the same use imposes costs on other people equal to MEC plus MUC. Therefore these costs should be subtracted from MDC to obtain the net value of consumption. What adjustment is then made to the accounts depends on the value originally included in them.

Although we have assumed that MDC is equal to the marginal willingness to pay for the resource, this may not be the case if the resource is subject to private ownership and if the owners take a long-term view of the profitability of the resource. In that case, some or all of the MUC may be included in the price and the above propositions have to be altered appropriately.

MOC and Optimal Resource Stocks

The requirement of sustainable use for a resource does not imply a particular stock level for that resource. The choice of the optimal level of stock is arrived at by comparing MOC and the marginal benefits at different stock levels, assuming that the rate of extraction is equal to the rate of regeneration, that is, the resource is being used on a sustainable basis. Because we are dealing with sustainable use only, MOC will be calculated as the sum of MDC and MEC at different stock levels. Equating MOC and the marginal benefits then defines the optimal stock level, which we call S^*.

It is quite possible that this optimal stock is different from the current stock level, S^0. For example, a resource-rich country may take the view that it should reduce its stock level permanently and use the proceeds to build up its productive capital. The path that the economy should take in going from S^0 to S^* is determined by a dynamic optimization exercise, in which the rates of capital accumulation and resource utilization are the key variables. The speed at which one proceeds will depend, among other things, on the marginal benefits of different levels of extraction, the marginal costs of those levels (MDC plus MEC), and the marginal productivity of capital. The scarcity premium associated with the use of the renewable resource is now endogenous to the whole analysis and does not have to be fed in as a separate piece of information. From the optimization, however, this value of MUC for the path from S^0 to S^* will be a potential output. Thus we require information on MDC and MEC to determine both the equilibrium stock level and the path by which we get to that stock level.

The Social Incidence of MOC

The present and future costs of resource exploitation fall on many parties, of whom some are significant users of the resource and others are not. From the point of view of policymakers, the incidence of these costs by income group is clearly important. Indications are that it is often the poorest people in the community who suffer the external costs of NRD. Although MOC does not provide the required information directly, the process of collecting the relevant data is facilitated when the framework for the estimation of MEC and MUC has been laid out.

Conclusion

The concept of marginal opportunity cost is not new, but in the context of nonsustainable use of renewable resources MOC functions as an organizing concept. The component parts of MOC focus attention on the relationship between resource depletion and its impacts elsewhere in the economy now and in the future. Moreover, MOC is linked to a view of the development process that emphasizes the role of renewable natural resources and argues that development and environmental preservation are inseparable parts of the process of social improvement.

The informational requirements for the calculation of MOC are considerable and, to some extent, subjective. This is because expectations have to be formed about the likely pattern of future exploitation and the likely future demands and supplies of the resource and its substitutes. Experience indicates, however, that the exercise can be undertaken and that the results, although necessarily approximate, are a useful tool in the planning and management of natural resources.

Notes

1. Such a path is very likely to have the features of Conway's (1983) concept of sustainability, that is, development will take place through diversification of inputs and outputs rather than monocultural activity.

2. For a detailed example of an evaluation that shows high rates of return to investments in reducing MOC from desertification, see Anderson (1987).

References

Anderson, Dennis. 1987. *The Economics of Afforestation: A Case Study in Africa.* Baltimore, Md.: Johns Hopkins University Press.

Ayres, Robert U., and Allen V. Kneese. 1969. "Production, Consumption and Externalities." *American Economic Review* 59.

Bartelmus, Peter. 1986. *Environment and Development.* London: Allen and Unwin.

Boulding, Kenneth. 1966. "The Economics of Coming Spaceship Earth." In *Environmental Quality in Growing Economy,* edited by H. Jarrett. Baltimore, Md.: Johns Hopkins University Press.

Brown, Lester. 1987. *Building a Sustainable Society.* New York: Norton.

Chambers, Robert. 1983. *Rural Development: Putting the Last First.* London: Longman.

Common, Michael. 1988. "Poverty and Progress Revisited." In *Economics, Growth and Sustainable Environments,* edited by D. Collard, D. W. Pearce, and D. Ulph. London: Macmillan.

Conway, Gordon. 1983. *Agroecosystem Analysis.* ICCET Series E, no. 1. Centre for Environmental Technology, Imperial College, London.

Hacche, Graham. 1979. *The Theory of Economic Growth.* London: Macmillan.

Hafkamp, Wim. 1984. *Economic-Environmental Modeling in a National-Regional System.* Amsterdam: North Holland.

Holdgate, Martin. 1982. *The World Environment, 1972–82.* Dublin: Tycooly.

International Union for Conservation of Nature and Natural Resources (IUCN). *World Conservation Strategy.* Gland, Switzerland.

Jones, Hywel. 1975. *An Introduction to Modern Theories of Economic Growth.* London: Nelson.

Kneese, Allen V., Robert U. Ayres, and Ralph d'Arge. 1970. *Economics and the Environment: A Materials Balance Approach.* Washington, D.C.: Resources for the Future.

Little, I. M. D., and J. A. Mirrlees. 1974. *Project Appraisal and Planning for Development Countries.* London: Heinemann.

Munasinghe, Mohan, and Gunter Schramm. 1983. *Energy Economics, Demand Management and Conservation Policy.* New York: Van Nostrand.

Norgaard, Richard. 1984. "Coevolutionary Development Potential." *Land Economics* 60.

———. 1987. "The Epistemological Basis for Agroecology." In Miguel A. Altieri and Richard B. Norgaard, *Agroecology: The Scientific Basis of Alternative Agriculture.* Boulder, Colo.: Westview Press.

Page, Talbot. 1977. *Conservation and Economic Efficiency.* Baltimore, Md.: Johns Hopkins University Press.

Pearce, David W. 1986. "The Limits of Cost Benefit Analysis as a Guide to Environmental Policy." *Kyklos* 1.

———. 1987. "The Economics of Natural Resource Degradation in Developing Countries." In *Sustainable Environmental Management: Principles and Practice,* edited by R. K. Turner. London: Belhaven Press, and Boulder, Colo.: Westview Press.

———. 1987. "The Foundations of an Ecological Economics." *Ecological Modelling* 38.

———. 1988a. "Optimal Prices for Sustainable Development." In *Economics, Growth and Sustainable Environments,* edited by D. Collard, D. W. Pearce, and D. Ulph. London: Macmillan.

———. 1988b. "Sustainable Development: Ecology and Economic Progress." University College London. Processed.

Pearce, David W., and Anil Markandya. 1987. "The Costs of Natural Resource Depletion in Low Income Developing Countries." University College London. Processed.

Repetto, Robert. 1986a. *Economic Policy Reform for Natural Resource Conservation.* Washington, D.C.: World Resources Institute.

———, editor. 1986b. *The Global Possible.* New Haven, Conn.: Yale University Press.

———. 1986c. *Natural Resource Accounting in a Resource Based Economy: An Indonesian Case Study.* Washington, D.C.: World Resources Institute.

———. 1986d. *World Enough and Time.* New Haven, Conn.: Yale University Press.

Sfeir-Younis, Alfredo. 1986. "Soil Conservation in Developing Countries." Western Africa Projects Department, World Bank, Washington, D.C. Processed.

Squire, Lyn, and Herman G. van der Tak. 1975. *Economic Analysis of Projects.* Baltimore, Md.: Johns Hopkins University Press.

Warford, Jeremy. 1987. *Environment and Development: Implementing the World Bank's New Policies.* Development Committee Pamphlet 17. World Bank, Washington, D.C.

Wilkinson, R. G. 1973. *Poverty and Progress.* London: Methuen.

World Health Organization. 1986. *World Health Statistics: Safe Water Supply and Sanitation: Prerequisites for Health for All.* Geneva.

World Resources Institute. 1986. *World Resources Report.* Washington, D.C.

5

The Environmental Basis
of Sustainable Development

Norman Myers

Degradation and destruction of environmental systems and natural resources are now assuming massive proportions in some developing countries and are a threat to continued, sustainable development. It is now generally recognized that economic development can be an important contributing factor to growing environmental problems in the absence of appropriate safeguards. A greatly improved understanding of the natural resource base and environmental systems that support national economies is needed if patterns of development that are sustainable can be determined and recommended to governments.

As the ultimate support of much economic activity, the environmental resource base makes a critical contribution to the cause of sustainable development. Especially in developing countries, environmental resources are increasingly being depleted (soil is being eroded, forests eliminated, and grasslands overgrazed) to a degree that adversely affects the prospects for sustainable development. There is an urgent need for policymakers to be supplied with an analytical framework for the problem so that they can systematically evaluate the tradeoffs involved and determine the most efficient points for policy interventions. These considerations apply more to developing countries than to developed countries, because developing countries are generally primary producers with large subsistence sectors and thus are more dependent on their natural resources, notably land and water.

This chapter addresses these issues, with emphasis on linkages, both environmental and economic, among various categories of natural resource systems and development sectors. It thus highlights certain spillover effects, principally in the form of lateral and temporal externalities (see also Myers 1986).

This is a revised and expanded version of a paper prepared for a Seminar on Land and Water Resources Management, Economic Development Institute of the World Bank, November 10–21, 1986, Washington, D.C.

The key questions are: What sorts of resource depletion are under way? What is their scale and scope? What types and levels of depletion are significant, serious, critical, or intolerable? Are there thresholds of irreversible injury? What are some forms of interconnection among various categories of resource depletion? How and to what extent does resource depletion generate adverse impacts on sectors, such as food production and public health, that are important for economic development?

Illustration of Concepts and Linkages: Deforestation

The issue of deforestation illustrates the character and extent of resource depletion and reveals some of the complex linkages at work. Tropical forests not only produce wood, whether commercial timber or fuelwood for subsistence. They also protect soils, retain moisture, and offer all manner of other environmental services. When forests are unduly degraded or destroyed—whether through overheavy logging, fuelwood gathering, or clearing for agriculture—the loss almost invariably extends far beyond the elimination of a source of wood. It extends to the productive capacity of the soil, which can be rapidly diminished by leaching, laterization, desiccation, and erosion. Sometimes soil fertility is irremediably reduced—the time required to make good the damage is longer than local farmers can wait, or the costs of technical remedies such as fertilizers are greater than farmers can sustain.

Further serious repercussions arise in areas far removed from the site of deforestation. Downstream, river systems have water flows that are excessively high during the rainy season and unusually low during the dry season. These flood-drought cycles now characterize many river systems below deforested watersheds, notably in southern Asia. The adverse consequences affect not only flood-prone lands but also irrigation-dependent croplands. Furthermore, soil and other debris from erosion causes downstream river beds to build up, aggravating the risk of flooding. It also leads to the sedimentation and siltation of irrigation channels, reservoirs, natural lakes, harbors, and offshore waters. In turn, these factors impinge upon the operation of irrigation systems, hydropower installations, domestic water supplies, port facilities, and fisheries, both inland and on coasts.

Other unwanted consequences of deforestation arise through linkages that are less direct and more diffuse. As forests are "mined" for fuelwood and as potentially renewable resources are harvested to effective extinction, rural households start to divert animal manure and crop residues from farm fields to house hearths. Fertilizer benefits give way to fuel needs, even though cropland productivity is further reduced (see chapter 8). It is estimated (Spears and Ayensu 1986) that in the developing countries of Asia and Africa alone, at least 400 million metric tons of dung are burned each year, and each ton means a loss of at least 50 kilograms of potential grain grown, or a total of 20 million tons of grain per year (some analysts calculate the total at twice as much). To put this calculation in perspective, 20 million tons of grain can feed 100 million people for one year. To purchase the grain on world markets could cost as much as US$3 billion—a figure to bear in mind when we consider the cost of establishing fuelwood plantations, which has been variously estimated at between US$900 million and US$1.3 billion a year for ten years (World Resources Institute and World Bank 1985).

Thus the use, or rather the misuse and overuse, of the total forest resource generates a backlash on other natural resources, including soils, water, hydropower potential, fish stocks, and gene reservoirs. Shortcomings in the forestry sector spill over into agriculture, energy, public health, communications, and fisheries, among other development sectors. These lateral externalities are paralleled by temporal externalities. Present use degrades the natural resource base to the detriment of the current generation in subsequent years and of generations into the indefinite future (for further clarification, see Myers 1986).

Central Issue: The Environment as a Sector of Development

Deforestation illustrates the many interdependencies between the natural world and the world of human beings. Our approach to natural resource systems should not be perceived in terms of restraints on development (an approach that ostensibly diverts funds and effort away from the goal of economic growth). Rather, we should regard environment as a valid sector of development or as an overarching sector that addresses the dynamic interactions among other sectors. Furthermore, the key question is no longer, "How can we best safeguard the environment?" Now it is, "How can we best make productive use, indeed expanded use, of our natural resources in order to further human welfare now and forever?"

Further Illustrations: Soil and Water

Soil

The earth's covering of soil is a crucial factor in agriculture, but in many parts of the developing world soil is being eroded and otherwise degraded at rates high enough to limit agricultural productivity. Erosion may cause rainfed crop productivity to decline by as much as 19 to 29 percent during the next twenty-five years (Food and Agriculture Organization 1984). The amount of agricultural land now being lost outright through soil erosion is estimated at a minimum of 20 million hectares per year; from 1985 to 2000, losses may reach a cumulative total of 540 million hectares (Sfeir-Younis 1986; see also Brown and Wolf 1982; El-Swaify and others 1985; and Pimentel and others 1985).

Hundreds of years are required to renew a mere 25 millimeters of soil, or the equivalent of 400 tons of soil per hectare (Hall and others 1982; Hudson, 1981). Yet an erosion rate of fifty tons per hectare per year is all too common in developing countries. The loss can be made good only by using increased amounts of chemical fertilizer. This heroic use of technology soon leads to declining returns: nitrogenous fertilizer put on eroded soil is often only one-third as effective as it is on uneroded soil (Lal 1983). At the same time, soil erosion is often difficult to discern. A loss of fifty tons per hectare per year amounts to only three millimeters of loss from the top of the soil, but it is enough to cause the productivity of most soils to plunge.

Where is erosion the worst? There is estimated to be an annual loss of 100 tons of soil on steep slopes in the Andes Mountains, in the middle reaches of the Yellow River basin in China, and in the black soil sectors of the Indian Deccan; and 200 to 500 tons in some gullied localities of Nepal. In fourteen large river basins of the developing tropics, which total 3.7 million square kilometers—an area almost equivalent to India and Pakistan together—erosion rates surpass 100 tons per hectare per year. In eleven of these basins it surpasses 200 tons, and in three, 500 tons. An average rate of soil loss, calculated by Sfeir-Younis (1986) for developing countries as a whole is fifty-three tons per hectare per year. For Africa the average figure is only 6 tons, for Latin America 12 tons, and for Asia 138 tons. Of course, a low average for Latin America overall is of scant interest to a farmer in the Acelhuate River basin of El Salvador, where the rate of soil loss is likely to approach 200 tons per hectare per year (Wiggins 1981).

Some of the worst erosion is in India. Land degradation of all forms, principally soil erosion, affects almost 1.75 million square kilometers of the country's 3.3 million square kilometers of territory, and 60 percent of the croplands suffer significant erosion. Overall, it has been estimated that the country's croplands are losing 6 million tons of soil per year, much if not most of it going eventually into the sea (Indian Council of Agricultural Research 1984; Narayana and Sastry 1985). In China, total soil loss can be roughly estimated at 4.3 billion tons per year from some 100 million hectares of croplands (Brown and Wolf 1982). In Ethiopia, the cultivated highlands are believed to be losing at least 1 billion tons of soil per year, and conceivably three times that amount (Hurni 1985; Lamb and Mylas 1983).

The main direct impact of soil erosion lies with decline in crop productivity (Follett and Stewart 1985; Lal 1983; Rijsberman and Wolman 1984; Schmidt and others 1982). Using the figure of 1 billion tons of soil lost per year in Ethiopia, we calculate that erosion accounts for a loss of crop output of at least 1 million tons of grain, which is equivalent to two-thirds of all the relief food shipped to Ethiopia in 1985. In India, soil erosion is estimated to cost farmers some 8.4 million tons of nitrogen, phosphorus, potash, and other critical soil nutrients each year. These nutrients are worth about US$6 million in terms of chemical-fertilizer replacements (Sfeir-Younis 1986). In general, erosion of good soils in the tropics results in maize-yield reductions of between 10 and 30 percent, while erosion of shallow or impoverished soils leads to reductions of between 30 and 70 percent and occasionally even more (Langdale and Schrader 1982). In parts of Mexico, maize yields have been reduced through severe erosion from 3.8 tons to only 0.6 tons per hectare (Sancholuz 1984), and in parts of Nigeria the decline is similarly pronounced, from 6.5 tons to only 1 ton per hectare (Lal 1983).

The second main impact of soil erosion is indirect: the sedimentation of hydropower facilities. According to a recent review (World Bank 1984) of development programs in 100 developing countries for 1982 to 1995, hydropower is projected to account for 43 percent of electricity production by 1995, at an investment cost of about US$10 billion. But if we look at the effects of sedimentation in 200 major dams built from 1940 onward, we find that a 2 percent sedimentation rate—a figure that can be used as a constant average—means that the live storage capacity of these dams will be reduced by one-third by the year 2000. Merely a 1 percent reduction in capacity will mean a loss of some 148,000 gigawatt hours of electricity by the year 2000. To produce an equivalent amount of electricity by thermal means would require 37 million tons of oil; and at US$80 per ton, or US$12 per barrel, sedimentation would constitute a direct cost of US$3 billion in electricity output forgone in the single year 2000. By way of comparison, US$3 billion is equal to around one-tenth of the development assistance provided in 1984 by the Organisation for Economic Co-operation and Development (OECD) and the Organization of Petroleum Exporting Countries (OPEC) (Pearce 1985).

Water

Unlike soil, water is an eminently renewable resource. Its stocks are available for recharge, either through natural hydrologic cycles or through human intervention. Yet all too often, it is used as a nonrenewable resource (Ambroggi 1980; Balchin 1985; Falkenmark 1984; Lunqvist and others 1985). Worse, water is rarely available in the right amount at the right place at the right time, largely because of precipitation patterns. Moreover, because of rising standards of living, demand for water in several parts of the world is increasing several times more rapidly than are population numbers.

As a measure of the imbalance between water needs and supplies, in the public health sector alone there are now 135 million more people in the developing world who do not have access to clean water or adequate sanitation than there were in 1970. This is especially significant for water-related diseases, which account for 80 percent of all sickness in developing countries and for 90 percent of the 15 million child deaths each year. Until this rate of child mortality can be reduced, there is little hope for family planning programs and their implications for population growth and development. Yet the megascale budget for the United Nations Water and Sanitation Decade, US$300 billion over a period of ten years, makes no provision for safeguarding water supplies at their principal source in forested catchments.

Water shortages affect agriculture even more. Irrigated croplands, which now constitute about 15 percent of all arable lands while producing 30 percent of all food, also account for 65 percent of the water used worldwide. More than half of all irrigated lands are in developing countries. Irrigation agriculture uses seven times more water than domestic needs and industry—the next two water-demanding activities—put together. Yet in several of the better irrigated parts of Asia, notably certain sectors of the Indian subcontinent, as well as Indonesia, the Philippines, and Thailand, the Green Revolution has lost some of its momentum as farmers find they can no longer rely on acceptable flows of irrigation water for their multiple crops of bumper-harvest rice each year (Cool 1984; Jayal 1984). In the Ganges River system, dry-season water flows declined by almost one-fifth during the 1970s (Bahugna 1978). Much of India's achievement in attaining food self-sufficiency has been due to a doubling of irrigated area since 1960. Yet, because of deforestation and the surging demand for domestic needs and industry (as well as sheer rise in human numbers), India faces the prospect of an acute water shortage in much of its territory by 1995 (Agarwal and Narain 1985). Despite the exceptional need for irrigation water, however, the superscale agricultural programs directed toward expanding irrigation pay little heed to the ultimate source of many water supplies, namely forested catchments (Gasser 1981).

Ten percent of the developing world's farmers live within the upstream catchments of rivers. This small segment of population thereby exercises much influence over the 40 percent of the farmers who live in valleys and floodplains (see Haigh 1982, an annotated bibliography; Hamilton 1983; and National Commission on Floods 1980). Not only does forest clearing lead to larger water flows, it also aggravates soil erosion, which causes the aggradation of stream and river beds and then disrupts water courses and increases flooding.

The worst flooding occurs on the Indian subcontinent, especially in India. The floodplains of India's rivers include 1.25 million square kilometers (38 percent of the country), and support more than 300 million people. Flood-prone lands have doubled in extent, from 20 million hectares in 1970 to 40 million in 1980 (Agarwal and Narain 1985; Dewan and Sharma 1985). In the Ganges Valley alone, the damage to crops, housing, public utilities, and other property now averages more than US$1 billion a year. Conversely, the average annual outlay on flood control works during the same period amounts to less than US$250 million: expenditures on watershed rehabilitation amount to even less.

Problems and Symptoms

In summary, we see a variety of emerging problems. Resource depletion is common in developing countries, and there are dynamic interactions among the various processes at work. The impacts of depletion spill over into several development sectors. Yet these are not so much problems as symptoms of deeper underlying problems.

Perception

Nature works as a unitary whole, within a seamless web of ecological interactions, and does not recognize the artificially imposed divisions of humans. It is therefore sometimes difficult to perceive the linkages that operate among natural resource systems. We do not readily comprehend the ecological repercussions of deforestation that arise through, for instance, overharvesting of fuelwood. As a result, we do not view activities such as tree planting with the same urgency that we view a "real priority" such as crop growing: people without food are considered to be worse off than people without fuelwood. Yet the two problems are not so disparate: deforestation leads to adverse consequences for agriculture. Fuelwood shortages affect other sectors such as public health: uncooked food is generally less nutritious and can contain pathogens that cooking would eliminate.

In short, we often lack the scientific and intellectual framework that would enable us to comprehend the multiple linkages at work. And once we have mastered these linkages we face the problem of tradeoffs—the paradoxes and dilemmas of apparently conflicting interests, some of which are hard to evaluate and quantify in a form that makes them readily comparable.

Asymmetry of Evaluation

Many of the natural resources in question—soil, water, and forests—are not usually priced in the marketplace. Although they are often exceptionally valuable, their value to people is not communicated by price signals. Conversely, the misuse and overuse of resources to generate marketable goods supply sensitive signals of their commercial worth. Thus, there arises an asymmetry of evaluation.

Fortunately, and by dint of innovative modes of analysis of these nonmarket outputs, we can come up with working estimates of some competing values. True, these proxy modes of analysis may appear crude in comparison with the refined modes of evaluation for conventional goods. Nonetheless, they serve to illuminate a complex situation. In Tanzania, for instance, it has been estimated (Peskin 1984) that deforestation leads to both environmental and economic externalities that, although ignored in conventional calculations of GNP, reduce net national product by 11 percent a year. Though the approach is less than rigorously scientific, it serves as a first-order approximation of the values of natural resources.

Ultimate Sources of Problems

Slash-and-burn agriculture accounts for more deforestation than all other agents put together (Food and Agriculture Organization and United Nations Environment Programme 1982; Myers 1984). Farmers who practice shifting cultivation are usually forced into this destruction by pressures over which they have little understanding or control. These pressures are generated by factors such as the maldistribution of land in established farming areas and lack of access to agronomic technology and credit systems. The farmers are no more to be blamed for felling the forest than soldiers can be held responsible for starting a war.

So we must be careful to distinguish between the proximate and ultimate causes of deforestation. By extension, the main response to deforestation no longer lies with measures that operate within the forest (nor are foresters the only, or indeed the best, persons to tackle the problem). The main response generally lies in areas far removed from the forest, wherever there is scope, for example, for intensified agriculture that can relieve the incentive for otherwise landless farmers to migrate into forests. By far the most productive way for Brazil to stem much of the spontaneous and unsustainable settlement of Amazonia is to engage in land reform in territories in the southern parts of the nation.

Similar considerations apply to processes of desertification in the Sahel. Because of the spread of cash crops grown for export in several countries, subsistence farmers have found themselves steadily pushed out of the principal food-growing areas and into areas with infertile and friable soils—precisely the soils most liable to degradation and desiccation.

This phenomenon of marginalization also occurs in other communities and environments. In parts of Central America and the Andes, an array of socioeconomic and political factors push small-scale farmers to the sidelines of the development process. They are compelled to move out from the better farmlands and up onto steeply sloped terrain, where they cause much soil erosion (Blaikie 1985; Posner and McPherson 1982). In many other areas in the developing world, people are pushed into marginal environments where they cannot avoid damaging natural resources, notably land and water supplies. By virtue of their impoverished status they tend to be people who are unusually vulnerable to external effects of both environmental and socioeconomic systems. Because of their rudimentary agricultural practices, those people are unusually inclined to damage environments that are vulnerable to misuse.

Conclusions

The environmental interconnections of natural resource systems constitute an "objective reality" that, whatever its compelling character, is often in conflict with the compartmentalized approach of human institutions. As more people make greater demands on the natural resource base that ultimately sustains much economic activity in the developing world, and as developing economies become increasingly complex and integrated, we can expect these conflicts to become more numerous, more complex, and more acute. In response, we need to adopt a more integrative approach to natural resource issues. Our limited understanding of environmental interactions underscores the need for research to clarify the issues, and for monitoring of the depletive processes that undermine the very basis of sustainable development.

Nonetheless, we possess enough of a grasp of the situation to allow us to do much more than before. The key to the challenge is to engage in more rough and ready appraisals by drawing on information and analysis that is already available. In general we can determine how such aspects as the speed of land degradation (notably soil erosion, but also salinization and other factors), the rate of water use, and the advancement of deforestation. We can also consider the consequences of overuse for stocks of associated resources and for dependent economic sectors. In other words, we must work more with common-sense judgments, even though those judgments may be based on deficient data.

Above all, enhanced perception of a given situation, its processes, and its root causes is needed.

Having come to grips in principle with the challenge, we need to formulate systematized modes of analysis to enable us to confront the tradeoffs at issue. At present it is difficult to identify tradeoffs, let alone define, document, or evaluate them. Moreover, although environmental safeguards can be expensive, they should be matched against the positive payoffs, which can be still more substantial (see Repetto 1986 for an exposition on this subject). Similarly, although the costs of action can be high, they need to be measured against the concealed costs of inaction.

We face many difficult choices. Tree planting needs to be balanced against food production. Soil conservation, with its long-term benefits must be weighed against rural activities that yield more immediate advantage. The benefits of watershed management by upstream communities must balance the needs of larger, downstream communities.

Perplexing as the choices are for decisionmakers (development planners, political leaders, and policymakers), let us bear in mind that in the meantime choices are being made daily, with broadscale impact, albeit with limited understanding of all factors involved. They are being made by millions of cultivators, forest dwellers, and others who decide by force of circumstance. If implicit micro-choices are already being made, they should be complemented by explicitly made macro-choices; by design rather than by default.

References

Agarwal, Anil, and Sunita Narain, editors. *The State of India's Environment 1984-85: The Second Citizen's Report.* New Delhi: Center for Science and Environment.

Ambroggi, Walter. 1980. "Water." *Scientific American* 243:101–16.

Bahugna, S. L. 1978. *Himalayan Trauma, Forests, Faults, Floods.* New Delhi: Gandhi Peace Foundation.

Balchin, W. G. V. 1985. "Water—A World Problem." *International Journal of Environmental Studies* 25:141–48.

Blaikie, P. L. 1985. *Political Economy of Soil Erosion in Developing Countries.* London: Longman.

Brown, L. H., and E. C. Wolf. 1982. *Soil Erosion: Quiet Crisis in the World Economy.* Worldwatch Paper 62. Washington, D. C.: Worldwatch Institute.

Cool, J. C. 1984. *Factors Affecting Pressure on Mountain Resource Systems.* Kathmandu: Agriculture Development Council.

Dewan, M. L., and S. Sharma. 1985. *People's Participation as a Key to Himalayan Eco-System Development.* New Delhi: Center for Policy Research.

El-Swaify, S. A., W. C. Moldenhauer, and A. Lo, editors. 1985. *Soil Erosion and Conservation.* Ankeny, Iowa: Soil Conservation Society of America.

Falkenmark, Malin. 1984. "New Ecological Approach to the Water Cycle: Ticket to the Future." *Ambio* 13:152–60.

Follett, R. F., and B. A. Stewart, editors. 1985. *Soil Erosion and Crop Productivity.* Ankeny, Iowa: Soil Conservation Society of America.

Food and Agriculture Organization. 1984. *Protect and Produce: Soil Conservation for Development.* Rome.

Food and Agriculture Organization and United Nations Environment Programme. 1982. *Tropical Forest Resources.* Rome: Food and Agriculture Organization and Nairobi: United Nations Environment Programme.

Gasser, W. R. 1981. *Survey of Irrigation in Eight Asian Nations.* Foreign Agricultural Economic Report 165. U.S. Department of Agriculture, Economics and Statistics Service, Washington, D.C.

Haigh, M. J. 1982. *Soil Erosion and Soil Conservation Research in India: An Annotated Bibliography.* Oxford, U.K.: Oxford Polytechnic Institute.

Hall, G. F., R. B. Daniels, and J. F. Foss. 1982. "Rate of Soil Formation and Renewal Rates in the USA." In *Determinants of Soil Loss Tolerance.* Madison, Wisc.: American Society of Agronomy.

Hamilton, L. S., editor. 1983. *Forest and Watershed Development and Conservation in Asia and the Pacific.* Boulder, Colo.: Westview Press.

Hudson, Norman. 1981. *Soil Conservation.* 2d ed. Ithaca, N.Y.: Cornell University Press.

Hurni, H. 1985. *Erosion, Productivity, and Conservation Systems in Ethiopia.* Berne: Soil Conservation Research Project, University of Berne.

Indian Council of Agricultural Research. 1984. *The Ganges River System: Hydrologic Disruptions, Flooding, Threats, and Irrigated Agriculture.* New Delhi: Indian Council for Agricultural Research.

Jayal, N. D. 1984. "Destruction of Water Resources—The Most Critical Ecological Crisis of East Asia." Paper presented at the Sixteenth Technical Meeting of the International Union for Conservation of Nature and Natural Resources, Madrid, Spain. Planning Commission of India, New Delhi.

Lal, Ratan. 1983. "Erosion-Caused Productivity Decline in Soils of the Humid Tropics." *Soil Taxonomy News* 5:4–5, 18.

Lamb, Robert, and S. Milas. 1983. "Soil Erosion: Real Cause of the Ethiopian Famine." *Environmental Conservation* 10:157–59.

Langdale, G. W., and W. D. Schrader. 1982. "Soil Erosion Effects on Soil Productivity of Cultivated Cropland." In *Determinants of Soil Loss Tolerance.* Madison, Wisc.: American Society of Agronomy.

Lundqvist, Jan, Ulrik Lohm, and M. Falkenmark, editors. 1985. *Strategies for River Basin Management.* Hingham, Mass.: Kluwer Academic.

Myers, Norman. 1984. *The Primary Source.* New York and London: W. W. Norton.

———. 1986. "Natural Resource Systems and Human Exploitation Systems: Physiobiotic and Ecological Linkages." Environment Department, World Bank, Washington, D.C.

———. 1986. "Tropical Forests: Much More Than Stocks of Wood." Background paper for Seminar on Land and Water Resources Management, Economic Development Institute of the World Bank, Washington, D.C.

Narayana, D. V. V., and G. Sastry. 1985. "Soil Conservation in India." In El-Swaify, Moldenhauer, and Lo (1985).

National Commission on Floods, Government of India. 1980. *A Report: Emergent Problems, with Special Respect to Irrigation.* New Delhi.

Pearce, D. W. 1985. "The Major Consequences of Land and Water Mismanagement in Developing Countries." Environment Department, World Bank, Washington, D.C.

Peskin, H. M. 1984. *National Accounts and the Development Process: Illustration with Tanzania.* Washington, D.C.: Resources for the Future.

Pimentel, David, and others. 1985. "World Food Economy and the Soil Erosion Crisis." Manuscript, Cornell University, Ithaca, N.Y.

Posner, J. L., and M. S. McPherson. 1982. "Agriculture on the Steep Slopes of Tropical America: The Current Situation and Prospects." *World Development* 10:341–53.

Repetto, Robert. 1986. *World Enough and Time.* New Haven, Conn.: Yale University Press.

Rijsberman, F. R., and M. G. Wolman, editors. 1984. *Quantification of the Effect of Erosion on Soil Productivity in an International Context.* Delft, Netherlands: Delft Hydraulics Laboratory.

Sancholuz, Lois. 1984. "Land Degradation in Mexican Maize Fields." Ph.D. diss., University of British Columbia, Vancouver.

Schmidt, B. L., and others, editors. 1982. *Determinants of Soil Loss Tolerance.* Madison, Wisc.: American Society of Agronomy.

Sfeir-Younis, Alfredo. 1986. "Soil Conservation in Developing Countries." Western Africa Projects Department, World Bank, Washington, D.C. Processed.

Spears, John, and E. S. Ayensu. 1985. "Resources, Development, and the New Century: Forestry." In *The Global Possible: Resources, Development and the New Century,* edited by Robert Repetto. New Haven, Conn.: Yale University Press.

Wiggins, S. L. 1981. "The Economics of Soil Conservation in the Acelhuate River Basin, El Salvador." In *Soil Conservation: Problems and Prospects,* edited by R. P. C. Morgan. Chichester, U.K: John Wiley.

World Bank. 1984. *World Development Report 1984.* New York: Oxford University Press.

World Resources Institute, World Bank, and United Nations Development Programme. 1985. *Tropical Forests: A Call for Action.* Washington, D.C.

6

Economic Incentives for Sustainable Production

Robert Repetto

The serious degradation of natural resources in developing countries stems not primarily from large projects, but from the cumulative effects of many small agricultural operations that cannot be reached by environmental impact assessment or regulation (IIED and World Resources Institute 1986, 1987). Remedies, therefore, must include changes in economic policies and incentives to promote sustainable resource use by large and small enterprises and households, and to channel economic and demographic growth into activities that raise incomes while preserving important natural resources.

The Need for Incentive Reform

Some incentive problems arise from market failure. For example, people borrow against the future by destroying renewable resources because they lack options. Small farmers around the world plant subsistence crops on marginal soils, even though the cost in erosion is high (World Commission on Environment and Development 1987). They persist in using inappropriate technologies because they lack the knowledge and resources to adapt. They ignore future consequences because institutions deny them a secure stake in the future yield of the resources they exploit. Solving these problems demands changes in incentives, so that people respond appropriately to true costs and opportunities. Market failures must be corrected, a difficult problem even in highly developed societies.

Resource degradation also stems from market distortions. Numerous government policies not only fail to reflect the true opportunity cost of resource use, but also encourage more rapid and extensive degradation of soils, water, and biota than would market forces alone. Many current policies—including subsidies, taxes, and market interventions—artificially increase the profitability of activities that result in serious

Table 6-1. Index of Nominal and Real Protection Coefficients for Cereal and Export Crops in Selected African Countries, 1972–83
(1969–71 = 100)

Country	Cereals 1972–73 Nominal index	Real index	Cereals 1981–83 Nominal index	Real index	Export crops 1972–73 Nominal index	Real index	Export crops 1981–83 Nominal index	Real index
Cameroon	129	90	140	108	83	61	95	75
Côte d'Ivoire	140	98	119	87	92	66	99	71
Ethiopia	73	55	73	49	88	71	101	66
Kenya	115	94	115	98	101	83	98	84
Malawi	85	79	106	100	102	94	106	97
Mali	128	79	177	122	101	83	98	70
Niger	170	119	225	166	82	59	113	84
Nigeria	126	66	160	66	108	60	149	63
Senegal	109	79	104	89	83	60	75	64
Sierra Leone	104	95	184	143	101	93	92	68
Sudan	174	119	229	164	90	63	105	75
Tanzania	127	88	188	95	86	62	103	52
Zambia	107	93	146	125	97	84	93	80
All Sub-Saharan Africa	122	89	151	109	93	71	102	73

Note: The nominal index measures the change in the nominal protection coefficient with border prices converted into local currency at official exchange rates. Data for Ghana are not available.
Source: World Bank (1986), p. 86.

resource degradation. Changing these policies would often reduce economic losses and long-term environmental degradation. Typically, these changes would also reduce fiscal burdens on government and eliminate important sources of inequity within the economy.

Eliminating these market distortions has large payoffs. Changes promote both economic growth and environmental quality, and thus command broad support. Unless market distortions are eliminated, investments and other programs that seek to enhance and protect natural resources will have little chance of overall success—efforts will be swept away by the expanding pattern of unsustainable resource use. Usually it is simpler to administer price adjustments, tax rates, and other existing policy instruments than to construct entirely new institutions or regulatory systems to cope with problems of market failure. Thus, elimination of market distortions is an important and feasible early step toward better resource management.

Agricultural Output Prices

Governments intervene in many ways in agricultural markets. The broad effect in developing countries is to turn the internal terms of trade against agriculture (World Bank 1986). Depressing agricultural profitability in this way reduces the derived demand for farmland, labor, and other inputs that are not supported by government subsidies. Because farmland cannot be massively shifted into other uses, the policies keep land prices lower than they otherwise would be. Consequently, returns on investment in the development and conservation of farmland are depressed. Farmers are discouraged from leveling, terracing, draining, irrigating, or otherwise improving their land. The loss of land productivity through erosion, salinization, or nutrient depletion is less costly relative to other values in the economy. In general, depressed agricultural prices lower the farmers' incentives to practice soil conservation.

Of course, prices are not the only incentive to farmers. Security of tenure is vital if rural households are to consider such long-term investments as soil conservation works or tree plantations. Many countries have found that ensuring secure rights to land, improvements, and tree stocks induces significant increases in household investment in conservation projects (see National Research Council 1986 for recent studies on tenure issues).

Within the agricultural sector, differential rates of implicit taxation among commodities can strongly influence cropping patterns and land uses. Many countries severely discriminate against export crops relative to domestic food crops such as cereals, as shown for Sub-Saharan Africa in table 6-1. Although many environmentalists argue that overemphasis

Table 6-2. *Vegetal Cover Factors for Erosion in West Africa*

Factors	Representative annual soil loss[a]
Bare soil	1.0
Dense forest or culture with a thick straw mulch	0.001
Savanna and grassland, ungrazed	0.01
Forage and cover crops (late planted or with slow development)	
First year	0.3–0.8
Second year	0.1
Cover crops with rapid development	0.1
Maize, sorghum, millet	0.3–0.9
Rice (intensive culture, second cycle)	0.1–0.2
Cotton, tobacco (second cycle)	0.5
Groundnuts	0.4–0.8
Cassava (first year) and yams	0.2–0.8
Palms, coffee, cocoa, with cover crops	0.1–0.3

a. Measured per unit of erodability defined for a standard bare plot of soil.
Source: Roose (1977), p. 51.

on export crop production exacerbates soil degradation and ecological disturbance, their view is not valid as a general proposition. First, most developing countries discriminate against export crops. Second, export crops, with some exceptions such as groundnuts and cotton, tend to be less dangerous to soils than basic food crops. Many export crops grow on trees and bushes that provide continuous canopy cover and root structure: coffee, cocoa, rubber, palm oil, and bananas can be quite suitable for the hillsides where they are often grown. As table 6-2 illustrates, in West Africa, where tree and bush crops are grown with grasses as ground cover, erosion rates typically are two to three times less than the rates for areas where staple crops such as cassava, yams, maize, sorghum, and millet are grown. Established pasturage also results in relatively low erosion rates.

Differential agricultural taxation can have a substantial effect on cropping patterns and land uses. Although many heavily taxed crops are perennials, ample evidence shows that over time farmers respond strongly to differential incentives (Askari and Cummings 1976). They respond more to differentials among crops than to overall discrimination against agriculture, because it is easier to shift land and other resources from one crop to another than it is to withdraw from agriculture altogether (Bale and Lutz 1981:8–22). Evaluation of agricultural price policies should not be divorced from assessments of land capability and considerations of soil conservation.

Subsidies for Agricultural Inputs

Pesticides

The use of pesticides in agriculture poses serious health and ecological risks, especially in developing countries (International Organization of Consumers Unions 1985). Farmers, farmworkers, their families, and consumers are extensively exposed, either in the field, by using contaminated containers, or by consuming contaminated food. Acute poisonings are common, and little is known about the effects of chronic exposure on people with such common health problems as anemia, liver abnormalities because of parasitic diseases, or reproductive disorders. The effect of pesticides on the immune system may exacerbate health problems in populations in which infectious diseases are prevalent (Olson 1986:20–25). Intensive pesticide use also creates significant ecological problems. Fish in irrigated rice paddies, ponds, and canals have been destroyed. Throughout the world pest populations have resurged and new pests have emerged as pesticides have killed off their natural predators. More than 400 pests have become resistant to one or more chemicals, and the number is growing exponentially (Dover and Croft 1984).

Few governments in developing countries have been able to establish workable systems of regulation and enforcement, training of farmworkers, and public education to ensure safe and effective use of pesticides. In fact, many governments in developing countries provide heavy subsidies to farmers who buy pesticides (see Repetto 1985 for a more complete discussion). Table 6-3 shows that in a sample of nine developing countries, subsidies range from 15 to 90 percent of full retail cost, with a median of 44 percent. In large countries these subsidies cost the governments hundreds of millions of dollars per year, and the fiscal burden is growing. These policies were put into place in the early years of the Green Revolution to induce small farmers to adopt an unfamiliar technology; they continue fifteen or more years later, even though the technology is by now familiar and the bulk of the subsidies go to large commercial farmers. Few, if any, governments have seriously investigated whether these funds could be better spent in research, training, extension, or regulation to promote better pest management practices.

Rational pest management balances the risks of crop losses against the costs of pest control. Using excessive amounts of chemicals is as irrational for the farmer as using none, especially when excessive use induces pest resistance and creates new pest problems. Most experts advocate integrated pest management (IPM), which relies on a balance of biological and chemical controls along with changes in practices such

Table 6-3. *Estimated Average Rate and Value of Pesticide Subsidies*

Country	Percentage of full retail costs	Total value (millions of U.S. dollars)	Value per capita (U.S. dollars)
China	19	285	0.3
Colombia	44	69	2.5
Ecuador	41	14	1.7
Egypt	83	207	4.7
Ghana	67	20	1.7
Honduras	29	12	3.0
Indonesia	82	128	0.8
Pakistan	. . .	. . .	. . .
Senegal	89	4	0.7

. . . *Negligible.*
Source: World Resources Institute (1985).

as cropping patterns and the timing of irrigation (Dover 1985). Pesticides should be used only at key stages in the life cycle of pests or crops or when damage to crops reaches a predefined threshold. By lowering pesticide costs to farmers, subsidies artificially depress this threshold and encourage prophylactic applications. Subsidies also artificially lower the costs of chemical use relative to other control methods such as planting resistant varieties of crops, destroying infected plants, and altering planting dates. Thus, they distort on-farm operating decisions and undermine the very approaches promoted by agricultural agencies. Removal of these subsidies may often be an opportunity to obtain economic, fiscal, health, and ecological benefits.

Fertilizers

Similar issues arise from the provision of subsidies for chemical fertilizers, although the problems are less acute. The rapid growth of fertilizer use in developing countries, a fourfold increase per hectare since 1970, has contributed to higher yields. Many developing countries have subsidized chemical fertilizers heavily since the 1960s. Even in the 1980s, according to the World Bank, "rates of subsidies . . . were rarely below 30 percent of delivered costs and were in some countries 80 to 90 percent (in Nigeria, for example). Rates of 50 to 60 percent are common" (World Bank 1986:95).

On close examination, many of the economic arguments for large, continuing subsidies are shaky (Berg 1986). After decades of experience, farmers should not need large subsidies to induce learning by doing or to overcome faulty perceptions of risk. If farmers are slow to adopt chemical fertilizers, it may be because of problems of distribution, extension, or availability of complementary inputs, not because of price. In parts of Africa and other regions of low population density, fallowing might be a more economical approach to restoring soil fertility. Fertilizer subsidies only partially offset explicit and implicit taxes on agricultural output, and are often captured by those who do not really need them (large commercial farmers of irrigated land) and those for whom they are not intended (producers and distributors).

Subsidies contribute to the inefficient use of fertilizer that is typical of developing countries. Imprecise timing and placement, careless use of irrigation water and other complementary inputs, and careless cultivation practices such as weeding contribute to application efficiencies that are probably well under 50 percent. Efficiency can be improved substantially at some additional costs of labor and management, but fertilizer subsidies distort these on-farm decisions. The result is the waste of costly inputs and increased pollution problems as chemicals run off into bodies of water.

More fundamentally, these subsidies artificially lower the cost of maintaining and restoring soil fertility and so reduce farmers' incentives to practice soil conservation. Loss of fertile topsoil and depletion of desirable properties in the soil can be offset to some extent by adding chemical fertilizers. If they are heavily subsidized, farmers do not realize the true costs of misusing their land.

Specifically, subsidies induce a substitution in favor of chemical fertilizers and against organic manures and crop residues. The amount of acreage under leguminous crops has fallen as the use of chemical fertilizer has expanded. In Taiwan, one of the few areas for which data are available, the use of organic manures and crop residues dropped from 17.3 million tons in 1962 to 7.1 million tons in 1981, and the acreage of green manure crops fell from 200,000 hectares in 1948–53 to only 18,000 hectares in 1981. Meanwhile chemical fertilizer use rose fivefold (Asian Productivity Center 1983).

Organic and chemical fertilizers are not perfect substitutes for one another. Although chemical fertilizers provide cheap, concentrated sources of certain nutrients, organic manures also provide a variety of micronutrients and improve soil structure. In sandy soils, they increase water retention and prevent nutrients from leaching out. They buffer soils against increases in acidity, alkalinity, and other toxicity. In clay soils, organic matter makes the soil more open and porous so that water infiltrates, thereby reducing runoff and erosion and preventing the baking and hardening of soil. Root development is improved, and biological activity is greatly stimulated. For these reasons, as numerous studies show, yields comparable to those produced by chemical fertilizers can be maintained through organic manuring. In addition, when used with chemical fertilizers, organic manures improve yields and offset the sharp declines in marginal returns to chemical fertilizers that most South and Southeast Asian countries have experienced (Kock 1985).

Heavy fertilizer subsidies have become an enormous fiscal burden with uncertain benefits and substantial environmental costs both on and off the farm. Because soil productivity is so vital to the development of most developing countries, these issues deserve more attention than they have received in the past.

Irrigation Water

At current prices, US$250 billion have already been invested in irrigation in developing countries, and US$100 billion more may be spent this century to create more capacity. The benefits in expanded farm output have been substantial. But there are serious economic and environmental problems, especially with large public systems (see Repetto 1986). Costs have been much higher and agricultural benefits lower than projected when investments were made (Faeth 1984). Operation and maintenance of completed systems are often deficient. Environmental impacts have been extensive. In India and Pakistan alone, over 20 million hectares have been lost through waterlogging, and at least 30 million are seriously affected by salinization (Jayal 1985; World Bank 1982).

Impounded water and canals provide breeding grounds and habitat for carriers of malaria and schistosomiasis. They have displaced whole communities and flooded valuable crop and forest lands, threatened critical ecosystems, and wiped out anadromous fish populations. The disruption of river hydrology downstream has caused erosion and sedimentation and had a great impact on estuaries and even deltaic fisheries (Pelts 1984).

The environmental and performance problems of irrigation systems are connected. More efficient use of water would reduce waterlogging, decrease the apparent need for additional, large-scale, increasingly costly expansions, and lessen the environmental effects of further river impoundment.

Irrigation is heavily subsidized, especially in public sector systems, and has become an enormous fiscal drain (see Repetto 1986). Revenues in most countries do not even cover operating and maintenance costs. Charges are also small relative to the value of water to farmers, especially during peak periods of need, so that excess demand is chronic and water has to be severely rationed. Farmers almost always desire additional water supplies because they bear few of the costs and enjoy most of the benefits.

This financing system undermines performance. Neither farmers, local governments, irrigation agencies, nor, for that matter, international banks are financially at risk for the success of investments in irrigation, so pressures for new capacity lead to a proliferation of projects, many of dubious worth. Benefit-cost analysis of such long-term investments is inherently speculative and easily becomes overly optimistic when strong political pressures are at work (Carruthers 1983).

These distortions tend to generate their own momentum. Areas that have not benefited from a heavily subsidized project also want one. Even in project areas, farmers in the head reaches, where canals are often finished first, are more favorably located to divert water from the reservoir, and they typically establish water-intensive cropping patterns and capture a disproportionate share of the available capacity. Because water is then chronically short in the tail reaches and is often less than what was promised to farmers there, strong demands for additional supplies are perpetuated (Small and others 1986).

Operation and maintenance of the system are also undermined by the excess demands generated by this method of financing. When funds for operation and maintenance depend on collections from irrigators, a vicious cycle of dissatisfaction, declining collections, and declining performance can ensue. When funds are allocated from general revenues, agencies are not accountable to users, and instead of providing an optimal service, they act as if allocating a limited resource (Bottrall 1981).

Operators are susceptible to pressure, inducement, and influence. And, when farmers' trust in the impartiality of the system is destroyed, they are less willing to abide by the system's rules or contribute to its upkeep. The fundamental problem lies in the financing system, which creates chronic excess demand and huge economic rents for those able to obtain water from public systems. These pressures would be much reduced if beneficiaries were financially responsible for the costs of the system, and users were charged approximately what water is worth to them.

Further, the current system generates few incentives to use water efficiently. For a few farmers, water is ample, cheap, and used rather lavishly. For most users, supplies are uncertain, irregular, and inadequate, which discourages complementary investments that would make the most effective use of the water supplied. Changes in financial incentives, in combination with management and physical improvements, would promote more efficient patterns of investment in irrigation, better operation and maintenance of existing systems, and more efficient use of water on the farm.

Mechanization

Developing countries promote agricultural mechanization through favorable taxes and tariffs, liberal allocations of rationed foreign exchange, cheap credit, and highly subsidized diesel fuel. Where governments take a large direct role in farming, parastatals use government budgetary resources to create highly mechanized operations and typically run them at a loss.

Such direct and indirect mechanization subsidies are at best unnecessary and at worst inefficient and inequitable. If it is economical and there is a reasonable supply system, farmers will use machinery without subsidies. If subsidies promote mechanization even when it is uneconomical, rural employment is reduced. Large landowners benefit, but smallholders with fragmented holdings derive little benefit and suffer a competitive disadvantage. Because smallholders derive considerable income from seasonal wage employment, they suffer further from labor displacement. Mechanization subsidies thus entail significant losses in economic welfare losses.

In addition, mechanization may severely damage natural resources. For example, using heavy equipment instead of traditional methods for land clearance in tropical regions has sometimes devastated the soil. Nutrients in the biomass have been lost, thin topsoils have been scraped off, the ground has been compacted so that water cannot infiltrate, and erosion rates have risen enormously. Even in less vulnerable areas, heavy equipment used inappropriately has compacted soils, reduced porosity, and increased susceptibility to erosion. When soils are left exposed and plowed against the contour, erosion is much worse than it would be under minimum tillage (National Research Council 1982:162ff). In general, eliminating subsidies for agricultural mechanization is another good example of a policy change that serves the complementary objectives of economic welfare and natural resource conservation.

Credit

Subsidized agricultural credit programs are at least as widespread in developing countries as subsidized fertilizers and other inputs and are even more questionable on economic grounds (Adams and others 1983). Their implications for natural resource management are not obvious but may be significant.

Special loan funds with interest ceilings, usually discounted by the lending institution with the monetary authorities, are often set up for purchasing particular inputs, growing particular crops, acquiring particular assets such as cattle or tractors, and developing land by clearing forests or constructing irrigation structures. In inflationary economies, real interest rates on these credits can be well below zero, and default rates are usually high because the lending institutions are largely absolved from risk.

These credit policies undermine financial institutions that serve rural areas; they cannot offer deposit rates high enough to attract rural savings. Their institutional capabilities are subverted by credit rationing, their ability to pass risk along to the monetary authorities, and freedom from competition. Inevitably, subsidized credit schemes in rural areas, even those specifically designed for smallholders, are quickly captured by larger farmers, who are considered to be better risks, more influential, and less costly to serve. The distribution of subsidized rural credit is typically even more skewed than the distribution of land.

Because credit is fungible and lenders cannot readily ensure that directed credit actually increases the flow of resources to the activities they intend to subsidize, the allocational effects are less clear-cut—and so are the impacts on natural resources. Only when loans are tied to verifiable activities, such as the acquisition of specific assets that might otherwise have been marginally attractive, are credit subsidies likely to have a significant impact on the allocation of resources. For example, several Latin American countries have highly subsidized credit for the acquisition of livestock and the establishment of ranches, often in forested regions. The economic prospects for some of these livestock operations have been risky, to say the least. Carrying capacity is low, and under minimal management much of the pasturage has within a few years lost its fertility and been invaded by weeds. Yet credit at negative real interest rates has increased demand for the assets (pasture land and cattle) that can be used as collateral, and this in turn has led to extensive deforestation in some countries.

The implications of credit subsidies on natural resource management have to be evaluated in each case. In general, to the extent that credit and machinery subsidies promote capital-intensive forms of agriculture with significant economies of scale, such as ranching, they displace farm labor (Branford and Glock 1985). Because rural populations are still growing in most developing countries and employment problems are acute, labor displacement puts marginal lands under even greater pressure.

Overview of Agricultural Input Subsidies

Subsidies have been backed up by government research, extension, and marketing services to promote a linear agricultural technology that draws heavily on natural resources for input and discharges wastes and residues as unwanted byproducts. This farming system diminishes soil productivity and the self-regulating capacities of agricultural ecosystems and compensates for these losses with chemical inputs. It imports large amounts of water and exports huge quantities of chemicals, minerals, and sediments in surface and underground runoff. Vast quantities of organic residues leave the farm sector as wastes and pollutants. With farm output doubling about every twenty years, this linear technology increasingly depletes and impairs natural resources.

Alternative agricultural systems rely more heavily on achieving a balance among the populations of various species, recycling nutrients, and sustaining productivity with a minimum of external inputs (Altieri 1983; Stonehouse 1981). Even in the United States and other industrial countries, where purchased inputs are relatively cheap, farms using alternative technologies are close to commercial viability and would probably be competitive if the external costs of chemical runoff and soil erosion were internalized into farm production costs. In the developing countries also, agricultural systems that use multiple cropping and integration of tree, animal, and crop production are capable of sustaining high productivity with fewer external inputs.

The current policy framework massively favors the dominant linear technology by heavily subsidizing inputs of chemicals, capital, and water and by failing to charge users of this technology the substantial external costs that exported residues and wastes imply. Even when an alternative approach would be more productive and stable in the long run, it is unlikely to emerge because policy-induced incentives are overwhelmingly biased against it. One of the first and most fundamental steps policymakers can take to promote more self-regenerative patterns of agricultural production is to reduce the pervasive bias against those patterns. If this step is not taken, special programs to promote more ecologically sound agricultural technologies are unlikely to make much headway in the field because their long-term economic value will not be apparent to the farmer.

Sectoral Issues

Forests

Every year more than 11 million hectares of forests are cleared for other uses, and in most developing countries deforestation is accelerating. In this century, the forested area in developing countries has fallen by half, with severe environmental consequences. In the tropics, forest clearance leaves only degraded soils that are unsuitable for sustained agricultural production. In watersheds, deforestation increases erosion, flooding, and sedimentation. In semi-arid areas, it robs the soil of essential organic matter and shelter from wind and water erosion. Moreover, in the tropics, loss of forest areas threatens the survival of uncounted species of animals and plants (World Resources Institution, the World Bank, and the United Nations Development Programme 1985).

There are also more direct economic losses. Although some forest conversion is to be expected, especially in richly endowed countries, the actual process of conversion has been highly wasteful, significant potential benefits to the local economy have been sacrificed, and the process has probably been pushed too fast and too far (Repetto and Gillis 1988). Rich public forests have been mined as though they were inexhaustible resources, and most of the proceeds have been needlessly relinquished to foreign interests. Forests have been opened for harvesting more rapidly and extensively than government forest agencies can manage and on terms that virtually ensured short-sighted, wasteful exploitation. Moreover, as the result of government-supported programs, huge forest areas have been sacrificed to cattle ranches, agricultural settlements, river impoundments, and other activities that have proven to be inferior uses of the land and invested capital or have failed outright.

The stumpage value, or economic rent, of mature virgin tropical forest timber is substantial. Many governments have offered timber concessions to logging companies on terms that capture only a small fraction of this rent in royalties, taxes, and fees, leaving most of it as above-normal profits for private interests. Table 6-4, which is derived from detailed country case studies, illustrates this fact. Governments typically lease timber lands not through competitive bidding, which would give them a larger share of the rents, but on the basis of standard terms or individually negotiated agreements. Potential investors are thus led to rush into concession agreements to grab the most favorable sites, setting off timber booms.

Other policies ensure rapid exploitation. Political instability, pressures from local partners, irregularities in the contracting process, and

Table 6-4. *Government Rent Capture in Tropical Timber Production*
(millions of U.S. dollars)

				Percentage share of government	
	Potential	*Actual*	*Official*		
Country and	*rent from*	*rent from*	*government*	*Actual*	*Potential*
period	*log harvest*[a]	*log harvest*[b]	*rent capture*[c]	*rent*	*rent*
Ghana, 1971–74	n.a.	n.a.	29	n.a.	n.a.
Indonesia, 1979–82	4,958	4,409	1,654	37.5	33.0
Philippines,					
1979–82	1,504	1,001	141	14.0	9.4
Sabah, 1979–82	2,065	2,064	1,703	82.5	82.5

n.a. Not available.

a. Potential rent assumes that all harvested logs are allocated to uses (direct export, sawmills, plymills) that yield the largest net economic rent.

b. Actual rent is the total rent arising from the actual disposal of harvested logs.

c. Rent capture is the total of timber royalties, export taxes, and other official fees and charges.

Source: Repetto (1988).

the risk that one-sided agreements will be renegotiated all lead concessionaires to realize their profits as early as possible. Host governments often require lessors to begin harvesting within a stipulated time and limit their leases to periods much shorter than a single forest rotation. Moreover, by basing royalties and taxes on the volume of timber harvested and not on the volume of merchantable timber present in the tract, governments encourage concessionaires to take only the trees of greatest value; in opening up a large forest area they damage or destroy many of the trees that are not deemed worth harvesting. The reform of forest revenue systems can conserve forest resources and increase the benefits to the host country.

Reforming incentives for local wood processing industries offers similar opportunities. Log-exporting countries have struggled to establish local processing industries and combat the discriminatory policies of industrial countries by waiving or reducing export taxes on processed wood and by banning log exports. These incentives to industrialization often promote local employment at a heavy environmental cost. Mills established in many countries in response to such inducements have been so small and inefficient that many more logs are needed to produce the same output as larger ones. Thereafter, governments are reluctant to reduce the mills' supplies of raw material, whatever economic or ecological reasons there might be to reduce the log harvest.

In forest-poor countries, woodlands, although nominally government property, are treated in practice as open-access, common property. Individuals lack adequate incentives to preserve tree stocks for future use or to plant and maintain new stocks that others might harvest. In addition, in many developing countries, individuals can still obtain title to forest lands by occupying and "improving" them, which means clearing at least part of the holding for agricultural or industrial use. Especially in Latin American countries, where existing agricultural lands are distributed very unequally, and in African countries, where population growth is extremely rapid, these laws and traditions obviously promote rapid deforestation.

Planting rates in many forest-poor countries would have to be increased many times over to balance sustainable yields with domestic demands. Yet prices for trees harvested from public lands are set far below their economic replacement cost. License fees, stumpage fees, and other charges for harvesting wood amount to only a minor fraction of the costs of planting new stocks and tending them to maturity. Higher fees might provide forest agencies with budgetary resources for forest maintenance and afforestation programs and encourage private farmers to establish woodlots on their holdings. There is no reason why forest products harvested from public lands in countries facing severe wood deficits should be sold at far less than their replacement costs.

Although current concerns about deforestation in the developing world have focused on population growth, land tenure, and investment needs, it is clear that government policies have been important factors behind resource depletion. Policy changes can do much to promote conservation and simultaneously raise the economic benefits countries glean from the forest sector.

Livestock

Most of the world's rangelands have deteriorated and are losing productivity often because of overstocking. Homogeneous herds selectively graze preferred grasses, expose and compact bare soil, and let less nutritious and palatable species take over. Water percolation decreases, soil erosion intensifies, water tables decline, and hardy shrubs replace grasses. Over 70 percent of the rangelands in developing countries are now moderately or severely desertified (IIED and World Resources Institute 1986:ch. 5).

Many governments and international agencies have supported range development and livestock services. Without adequate control over herd sizes, however, stocking rates have typically risen to exceed the carrying capacity of the range in years of low rainfall and forage production. Managing an intrinsically communal rangeland resource to limit overgrazing is critical. Various proposals have been made to finance infrastructure investments and livestock services through locally administered grazing fees and head taxes in order to discourage the excessive size of herds. These proposals have been difficult to reconcile with local political and cultural traditions.

Beyond support for livestock services, governments, especially in Latin America, have offered generous fiscal and financial support: subsidized credits, tax holidays and exemptions, and export incentives. Although pasturage is an appropriate land use over vast areas of Latin America, these inducements, together with the skewed distribution of landownership, have probably kept many large holdings in livestock production, which is a land- and capital-intensive operation, instead of crop production. As a result, there are fewer agricultural employment opportunities and greater pressure on smallholdings to increase cropping intensity to meet income and food requirements. On marginal lands, an increase in cropping intensity aggravates erosion and depletes fertility.

Moreover, generous fiscal inducements have led ranchers to convert large forested areas to extensive pasture in areas such as the Amazon. Under minimal management and without adequate fertilization, many converted soils lose nutrients through leaching and are invaded by weeds, so that within a few years productivity declines and the pasture is abandoned. Nonetheless, such operations may still be privately profitable because of the policy incentives to investors.

A recent analysis of large, government-supported ranches in the Brazilian Amazon showed, for example, that a typical ranch was intrinsically uneconomic because of low productivity and relatively high establishment costs; it could be expected to lose more than half of its invested capital over the lifetime of the project. Nonetheless, the investments were still highly profitable for private investors, whose own own equity tripled (Repetto 1987). In the late 1970s (and continuing to the present for approved projects), such investors could write off operating losses against other unrelated taxable income and were eligible for income tax holidays, accelerated depreciation, tax credits for investments in approved Amazonian projects, and subsidized credit at negative real interest rates.

Through forgone tax revenues and loan capital that could be repaid in inflated currency, the Brazilian government effectively financed a large share of approved livestock investments, affecting many millions of hectares of forest land. The government also bore a substantial share of operating losses. Private investors could shelter outside income by acquiring cattle ranches in the Amazon with very little equity investment and could take advantage of rising land prices for an ultimate capital gain. Such policies run the risk of promoting economically and ecologically unsound investments at substantial fiscal cost to the government. Policy change can help to ensure that livestock investments are made in regions and in technologies that offer a reasonable prospect of success, both economically and in their use of land and water resources.

Conclusion

Although environmental problems in development are widespread and serious, much can be accomplished by taking advantage of opportunities to establish policies that promote better resource management and conservation and, at the same time, reduce fiscal burdens on government to improve economic productivity. By and large, among countries heavily dependent on their natural resources for sustained income growth, there is no conflict between good resource management and sound development policy.

References

Altieri, Miguel A. 1983. "Agroecology: The Scientific Basis of Alternative Agriculture." Berkeley: University of California, Division of Biological Control.

Asian Productivity Center. 1983. *Recycling Organic Matter in Asia for Fertilizer Use*. Tokyo.

Askari, H., and T. J. Cummings. 1976. *Agricultural Supply Responses: A Survey of Econometric Evidence*. New York: Praeger.

Bale, Malcolm D., and Ernst Lutz. 1981. "Price Distortions in Agriculture and Their Effects: An International Comparison." *American Journal of Agricultural Economics* 63(1):8–22.

Berg, Elliot. 1986. "Economic Issues in Fertilizer Subsidies in Developing Countries." Background paper for the *World Development Report 1986*. World Bank, Washington, D.C..

Bottrall, Anthony F. 1981. *Comparative Study of the Management and Organization of Irrigation Projects*. World Bank Staff Working Paper 458. Washington, D.C.

Branford, Sue, and Oriel Glock. 1985. *The Last Frontier: Fighting over Land in the Amazon*. London: Zed Books.

Carruthers, I. 1983. *Aid for the Development of Irrigation*. Paris: Organisation for Economic Co-operation and Development.

Dover, M. 1985. *A Better Mousetrap: Improving Pest Management for Agriculture*. Washington, D.C.: World Resources Institute.

Dover, M., and B. Croft. 1984. *Getting Tough: Public Policy and the Management of Pesticide Resistance*. World Resources Institute, Washington, D.C.

Faeth, P. 1984. *Determinants of Irrigation Performance in Developing Countries*. Washington, D.C.: U.S. Department of Agriculture, Economic Research Service, International Economics Division.

IIED (International Institute for Environment and Development) and World Resources Institute. 1986. *World Resources 1986*. New York: Basic Books.

————. 1987. *World Resources 1987*. New York: Basic Books.

International Organization of Consumers Unions, Regional Office for Asia and the Pacific. 1985. *The Pesticide Poisoning Report*. Penang, Malaysia.

Jayal, N. D. 1985. "Emerging Patterns of the Crisis in Water Resource Conservation." In *India's Environment: Crisis and Response*, edited by J. Bandhyopadhyay. Dehra Dun: Natraj.

Kock, W. 1985. "Aspects of Agricultural Research on Rice and Rice-Based Farming Systems in Bangladesh." Paper prepared for the World Bank Resident Mission in Bangladesh, Dhaka, March 20.

National Research Council. 1982. *Ecological Aspects of Development in the Humid Tropics*. Washington, D.C.: National Academy Press.

National Research Council, Board on Science and Technology for International Development. 1986. *Proceedings of the Conference on Common Property Resource Management*. Washington, D.C.: National Academy Press.

Olson, L. J. 1986. "The Immune System and Pesticides." *Journal of Pesticide Reform* (Summer):20–25.

Pelts, G. E. 1984. *Impounded Rivers: Perspectives for Ecological Management*. New York: Wiley.

Repetto, Robert. 1985. *Paying the Price: Pesticide Subsidies in Developing Countries*. Washington, D.C.: World Resources Institute.

————. 1986. *Skimming the Water: Rent-Seeking and the Performance of Public Irrigation Systems*. Washington, D.C.: World Resources Institute.

————. 1987. "Creating Incentives for Substantial Forest Development." *Ambio* 16(2).

————. 1988. *The Forest for the Trees: Government Policies and the Misuse of Forest Resources*. Washington, D.C.: World Resources Institute.

Repetto, Robert, and M. Gillis, editors. 1988. *Public Policies and the Misuse of the World's Forest Resources*. Cambridge, U.K.: Cambridge University Press.

Roose, E. 1977. "Erosion et ruissellement en Afrique de l'Ouest: Vingt années de mesure, en petites parcelles expérimentales." Paris: ORSTROM/IITA.

Small, L., and others. 1986. "Regional Study on Irrigation Service Fees: Final Report." International Irrigation Management Institute, Kandy, Sri Lanka.

Stonehouse, B. 1981. *Biological Husbandry*. London: Butterworths.

Von Pischke, J. D., Dale W Adams, and Gordon Donald. 1983. *Rural Financial Markets in Developing Countries*. Baltimore, Md.: Johns Hopkins University Press.

World Bank. 1986a. *World Development Report 1982*. New York: Oxford University Press.

————. 1986b. *World Development Report 1986*. Part II, "Trade and Pricing Policies in World Agriculture." New York: Oxford University Press.

World Commission on Environment and Development. 1987. *Our Common Future*. Oxford, U.K.: Oxford University Press.

World Resources Institute. 1985. *Paying the Price: Pesticide Subsidies in Developing Countries*. Washington D.C.

World Resources Institute, World Bank, and United Nations Development Programme. 1985. *Tropical Forests: A Call for Action*. Washington, D.C.

7

Deforestation in Brazil's Amazon Region: Magnitude, Rate, and Causes

Dennis J. Mahar

The world's tropical rainforests are disappearing at an alarming rate. These forests, which once occupied 16 million square kilometers of the earth's surface, today cover only 9 million square kilometers. It is estimated that Latin America and Asia have already lost 40 percent of their original forests, and Africa a little more than half (Myers 1984). In many countries, the rate of deforestation is accelerating: most of the forested areas of Bangladesh, India, Philippines, Sri Lanka, and parts of Brazil, for example, could be gone by the end of this century. Only in the Congo Basin and in some of the more isolated areas of Amazonia does the forest remain largely intact.

Although deforestation of this magnitude has in some instances yielded considerable short-term benefits through timber exports and agricultural production on previously forested land, it has entailed huge (and largely unmeasured) long-term costs both to the people of the countries directly affected and to the earth as a whole. Among the more direct and visible of these costs are the losses of forest products such as timber, fuelwood, fibers, canes, resins, oils, pharmaceuticals, fruits, spices, and animal hides. More indirect, but equally important, long-term costs include soil erosion, flooding, and the siltation of reservoirs and hydroelectric facilities; destruction of wildlife habitat; climatic changes; and the irreversible loss of biological diversity.

According to Myers (1986), the main proximate causes of tropical deforestation worldwide are small-scale agriculture, commercial logging, fuelwood gathering, and cattle raising. The underlying causes, however, include poverty, unequal land distribution, low agricultural productivity, rapid population growth, and various public policies. The purpose of this chapter is to shed further light on the subject by analyzing the effects of certain government policies on deforestation in the Brazilian Amazon. The coverage is far from complete. The emphasis is on policies that encourage environmentally unsound economic activities, and on

Figure 7-1. The Amazon Region of Brazil

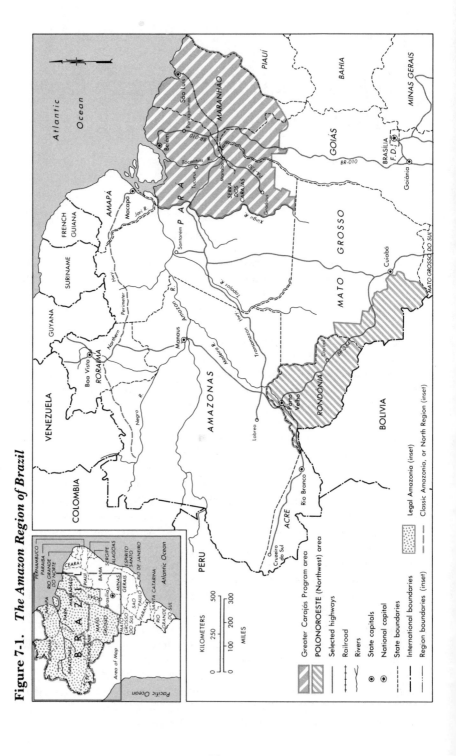

those subregions of Amazonia that are now experiencing the most rapid deforestation. The principal message is that attempts to reduce or stop tropical deforestation by fiat alone are not likely to succeed if economic incentives encourage people to do the opposite. Legislation to introduce land-use zoning, establish national parks, or prohibit certain economic activities, for example, must be accompanied by the removal of incentives that conflict with these goals.

Deforestation of the Brazilian Amazon

According to estimates of the Food and Agriculture Organization (FAO), Brazil contains about 3.5 million square kilometers of tropical forests (Guppy 1984:930). The area includes 30 percent of the world's total amount of tropical forest and is more than the combined forested areas of Colombia, Indonesia, Peru, and Zaire. Almost all of Brazil's standing tropical forests are located in the Amazon Basin, which has been characterized as the "single richest region of the tropical biome" (Myers 1984:50). As legally defined, the Brazilian Amazon region consists of nearly 5.0 million square kilometers, or 58 percent of the country's total land area (see figure 7-1).[1] About half of the region (2.5 million to 2.8 million square kilometers) is upland (*terra firme*) where the original vegetation was tropical rainforest (Fearnside 1986b). Perhaps another 500,000 square kilometers consists of transitional forests (U.S. Department of Energy 1986). There are in addition large areas of savanna (*cerrado*) in the southern reaches of legal Amazonia in the states of Mato Grosso and Goiás.

The first scientifically based estimates of deforestation in the Brazilian Amazon were made in the early 1970s by staff of the Brazilian government's Radar in Amazonia (RADAM) project, which employed airborne side-looking radar to gather primary data. These estimates suggested that relatively little clearing of the forest had taken place. More comprehensive estimates of deforestation were derived from Landsat satellite images a few years later. They indicated that roughly 30,000 square kilometers, or about 0.6 percent of Amazonia (approximately 1.0 percent of the forest), had been cleared as of 1975 (see table 7-1). When these early Landsat images were first made public, they were cited as proof that the environmentalists—some of whom had predicted the demise of the Amazonian forest by the end of the century—had greatly exaggerated their case (Denevan 1973). More recent data, however, make it clear that there was no cause for complacency.

Landsat images indicate that deforestation has accelerated sharply since the mid-1970s. As shown in table 7-1, the deforested area increased to 125,000 square kilometers by 1980 and to almost 600,000 square kilometers by 1988. The latter area is equivalent to 12 percent of

Table 7-1. Landsat Surveys of Forest Clearing in the Brazilian Amazon
(area in square kilometers)

State or territory	Area	Area cleared				Percentage of total area classified as clear			
		1975	1978	1980	1988	1975	1978	1980	1988
Acre	152,589	1,165.5	2,464.5	4,626.8	19,500.0	0.8	1.6	3.0	12.8
Amapá	140,276	152.5	170.5	183.7	571.5	0.1	0.1	0.1	0.4
Amazonas	1,567,125	779.5	1,785.8	3,102.2	105,790.0	0.1	0.1	0.2	6.8
Goiás	285,793	3,507.3	10,288.5	11,458.5	33,120.0	1.2	3.6	4.0	11.6
Maranhão	257,451	2,940.8	7,334.0	10,671.1	50,670.0	1.1	2.8	4.1	19.7
Mato Grosso	881,001	10,124.3	28,355.0	53,299.3	208,000.0	1.1	3.2	6.1	23.6
Pará	1,248,042	8,654.0	22,445.3	33,913.8	120,000.0	0.7	0.8	2.7	9.6
Rondônia	243,044	1,216.5	4,184.5	7,579.3	58,000.0	0.3	1.7	3.1	23.7
Roraíma	230,104	55.0	143.8	273.1	3,270.0	0.0	0.1	0.1	1.4
Legal Amazon	5,005,425	28,595.3	77,171.8	125,107.8	598,921.5	0.6	1.5	2.5	12.0

Sources: Fearnside (1986b) and World Bank estimates.

Amazonia and is larger than the entire country of France. Moreover, damage to the forest has been concentrated in certain subregions, not randomly distributed. In this regard, the situation in Mato Grosso and Rondônia—where nearly one-fourth of the forest has already been cleared—may be contrasted with the situation in Amapá where more than 99 percent of the forest is still intact. Although not fully revealed by the state-level data in table 7-1, particularly intense clearing has taken place along the region's major overland routes: the Belém-Brasília highway and its zone of influence in southern Pará and northern Goiás, and the Cuiabá-Porto Velho highway and its associated feeder roads in Mato Grosso and Rondônia (Fearnside 1986c; Woodwell, Houghton, and Stone 1986).

It is not possible to estimate with any degree of precision the relative contribution to deforestation of various economic activities. It seems evident, however, that the rapid expansion of Brazil's agricultural frontier over the past two decades has been the most important factor. According to agricultural census data, the area used for agricultural purposes in Amazonia increased from 313,000 square kilometers in 1970 to more than 900,000 square kilometers in 1985. This expansion occurred in virtually all of the region's states and territories. Consistent with the deforestation estimates derived from Landsat images, the census data show that the spread of agriculture was particularly rapid in northern Mato Grosso and Goiás, southern Pará, and Rondônia.

Details on agricultural land-use patterns for 1985 have not yet been published. The data summarized in table 7-2 indicate that approximately 145,000 square kilometers of the Brazilian Amazon were in agricultural use, either for crops or livestock, as of 1980. Pasture is clearly the predominant agricultural land use in the region, and cattle ranching therefore would appear to be the major proximate cause of deforestation. Based on comparisons with the 1970 census data, the conversion of forest to pasture occurred at the rate of approximately 8,000 to 10,000 square kilometers per year during the 1970s. Most, but not all, of this pasture formation took place on large landholdings.

Land devoted to annual cropping, the second most important form of agricultural land use, probably increased by about 2,000 square kilometers per year between 1970 and 1980. This is typically a small-farmer activity. Farm plots devoted to annual crops are frequently sold or abandoned after only a few years of use, however, as a result of rapidly declining yields. These areas are then converted to pasture—often by larger landowners—or are quickly invaded by an impoverished secondary growth known as *capoeira*. Given this traditional sequence of land use in Amazonia—from undisturbed forest to annual crops to pasture or secondary growth—it is likely that much of the deforestation attrib-

Table 7-2. *Agricultural Land Use in Amazonia, 1980*

Use	Area (square kilometers)	Percent
Crops		5.9
Annual[a]	42,231.6	
Perennial	7,619.5	
Subtotal	49,851.1	
Pasture	94,098.1	11.1
Undisturbed[b]	704,994.3	83.0
Total	848,943.5	100.0

Note: Amazonia is defined as the North region (Acre, Amapá, Amazonas, Pará, Rondônia, and Roraima) plus northern Mato Grosso and Goiás.

a. Includes fallow land.

b. Forest, natural pastures, and areas (such as rivers and mountains) unsuitable for agricultural use.

Source: IBGE (1983).

uted to livestock development has actually been caused by the spread of small-scale agriculture.

Logging has also grown rapidly in Amazonia over the past two decades. Between 1975 and 1985, regional roundwood production increased from 4.5 million cubic meters per year (14.3 percent of the national total) to 19.8 million cubic meters (46.2 percent of the national total). It is not clear, however, how much deforestation can be attributed to logging per se since much timber extraction in Amazonia is a by-product of land clearing for agricultural purposes. Loggers selectively cut only commercially valuable species in newly opened areas. Such trees usually represent a very small proportion of the standing forest—as of the late 1970s only five species (out of an estimated 1,500) accounted for 90 percent of the region's timber exports (Browder 1988). The vast majority of the trees, which are unknown in extraregional markets and thus have little or no commercial value, are burned before the planting of crops. Except for the Jari project in northern Pará, practically no replanting of trees is done in Amazonia.

The Role of Government Policies

Government policies designed to open up Amazonia for human settlement and to encourage certain types of economic activity have played a major role in the deforestation process. Massive road-building programs in the l960s and 1970s made large areas of the region accessible by overland means for the first time, while government-sponsored settlement schemes simultaneously attracted migrants from Brazil's Northeast and South regions. Special fiscal incentives and subsidized credit lines encouraged land uses such as cattle raising, which allowed a relatively small population to have a large impact on the rainforest. Greater details on the objectives, content, and results of these policies follow.

Operation Amazonia

As of 1960, Amazonia had only 2.5 million inhabitants and a per capita income just half what it had been in 1910 (Santos 1980). The military government that came to power in 1964 gave high priority to reversing the region's economic and demographic stagnation. In a series of legislative acts and decrees in 1966 and 1967 (cumulatively known as Operation Amazonia), the new government firmly committed itself to the development and occupation of the region, as well as the eventual integration of Amazonia with the rest of Brazil. These plans included an ambitious road-building program to link Amazonia with the Northeast and South, agricultural colonization schemes, and fiscal incentives for attracting new industrial and agricultural enterprises. An administrative structure, including a regional development agency (Superintendency for the Development of Amazonia, or SUDAM) and a regional development bank (Bank of Amazonia, or BASA), was created to coordinate the implementation of these plans.

The motives behind Operation Amazonia were to a large extent geopolitical. Several neighboring countries (particularly Peru and Venezuela) were already well advanced in programs to occupy and develop their respective Amazon regions, and Brazil's military leaders were anxious to ensure national sovereignty by establishing self-sustaining settlements in frontier areas. Because it was believed that vast quantities of natural resources remained hidden in the forest, this posture is understandable. But in the design and implementation of an economic development strategy little thought was given to the unique physical and human environments of Amazonia.

The Belém-Brasília Highway

Amazonia had only 6,000 kilometers of roads in 1960, of which less than 300 kilometers were paved. Except by air and long sea routes, the region was virtually cut off from the rest of Brazil. Intraregional travel was also difficult, and the sparse populations tended to cluster in the region's two major cities, Belém and Manaus, and in small towns and villages scattered along the Amazon River and its 1,100 tributaries. Although this centuries-old isolation from the more dynamic South had arguably retarded the region's economic development, it had also protected the rainforest from destruction. The physical isolation of Amazonia—and the protection this provided to the rainforest—came to an end in 1964 with the completion of a 1,900-kilometer-long, all-weather highway that connected the new capital city of Brasília in Brazil's heartland with Belém at the mouth of the Amazon River.

Large numbers of migrants in search of land and employment entered the region via the Belém-Brasília highway (BR-010). So did large firms that wished to establish cattle ranches to take advantage of the cheap land and generous tax and credit incentives offered by the government. Official estimates suggest that the total human population in the zone of influence of the highway (which includes some area outside of Amazonia) increased from 100,000 in 1960 to 2 million ten years later. There can be no doubt that the surge of migration and economic activity stimulated by the Belém-Brasília highway contributed to widespread deforestation. Around 1970, one traveler noted the devastation and apparent abandonment of the land on either side of the highway, and described the area around Paragominas in southern Pará (the site of numerous SUDAM-approved cattle ranches) as a "ghost landscape" (Paula 1971).

Environmental degradation was not confined to areas adjacent to the main highway. The increase in population associated with the Belém-Brasília highway quickly generated demand for secondary and feeder roads, which in turn attracted more population, and so on. Landsat photos vividly illustrate the impact on the rainforest of one such highway (PA-150) opened to traffic in the late 1960s. In southern Pará the road crosses 47,000 square kilometers (a small area by Amazonian standards, but about the size of Switzerland) in which the cleared area increased dramatically from 300 square kilometers in 1972 (0.6 percent of the area), to 1,700 square kilometers in 1977 (3.6 percent), to 8,200 square kilometers (17.3 percent) in 1985. Although the relative contribution of different activities to this deforestation is not known with precision, it is virtually certain that the conversion of forest to pasture has been a leading cause.

Incentives for Livestock Development

To attract private enterprise to the region, Operation Amazonia increased public expenditures on infrastructure—for example, roads, airports, telecommunications—and special fiscal incentives and credit lines for firms willing to establish operations in Amazonia. The package of fiscal benefits available to qualifying firms was extensive and included holidays from the corporate income tax for a period of ten to fifteen years, as well as exemptions from export taxes and import duties.

Investment tax credits. The most powerful of the incentives allowed registered Brazilian corporations to take up to a 50 percent credit against their federal income tax liabilities if the resulting savings were invested in projects located in Legal Amazonia and approved by SUDAM. Investment projects could be new enterprises or simply the expansion or modernization of existing enterprises. Under the 1963 legislation that created the investment tax credit, only industrial projects were eligible; in 1966 eligibility was expanded to include projects in the agricultural, livestock, and service sectors.[2] Depending on the priority assigned to a given project by SUDAM, tax-credit funds could constitute up to 75 percent of investment. Since 1979, the approval of livestock projects in the rainforest (*floresta densa*) has been officially prohibited, although it has been difficult to enforce this rule.

The tax-credit mechanism proved very attractive to investors and by late 1985 about 950 projects had been approved by SUDAM. Of this total, 631 projects were in the livestock sector (Garcia Gasques and Yokomizo, 1986: 51).[3] These livestock projects at present cover 8.4 million hectares, or an average of about 24,000 hectares (hectares) per ranch. SUDAM-approved livestock projects have now been established in all parts of legal Amazonia, although about three-fourths are in southern Pará and northern Mato Grosso. These projects have probably been the single most important source of deforestation in these two subregions. Their relative contribution to deforestation in Amazonia as a whole, however, has clearly been much smaller, probably less than 10 percent of the total.

Over the years, livestock projects have absorbed about 44 percent of the SUDAM-administered tax-credit funds. In absolute terms, total disbursements to the owners of these projects have amounted to the equivalent of approximately US$700 million (Browder 1987). Despite this huge subsidy, only ninety-two livestock projects have been awarded certificates of completion by SUDAM. Moreover, the performance of most of these completed projects has fallen far short of expectations. In a sample of nine such projects, selected as part of a field survey carried out by the Institute of Economic and Social Planning (IPEA), the average level of production was found to be less than 16 percent of that originally projected; three of the nine projects visited were not producing anything (Garcia Gasques and Yokomizo 1986:56).

The IPEA study attributes the poor performance of SUDAM-approved livestock projects largely to administrative and management problems such as inadequate purchases of breeding stock, frequent changes in project ownership, delays in the release of fiscal incentive funds, cost escalation, and weak supervision on the part of SUDAM. Surely these problems constitute part of the explanation. A recent study, however, argues convincingly that cattle ranching under conditions commonly prevailing in Amazonia is intrinsically uneconomic (Hecht, Norgaard, and Possio n.d.).

Table 7-3. *Internal Rates of Return to a Typical SUDAM-Approved Livestock Project under Two Scenarios*
(percent)

	Increase in land value		
Scenario	*0%*	*15%*	*30%*
High cattle prices			
Appropriate grazing intensity			
Corporation resources[a]	16	18	24
All resources[b]	−1	2	9
Overgrazing			
Corporation resources	23	24	27
All resources	−2	0	4
Low cattle prices			
Appropriate grazing intensity			
Corporation resources	−3	6	17
All resources	−14	−6	5
Overgrazing			
Corporation resources	16	18	23
All resources	−10	−7	−1

Note: Low input prices are assumed in both scenarios.
a. Ignores capital expenditures financed through fiscal incentives and official credit.
b. Fiscal incentives and official credit treated as if they were corporation's own capital.
Source: Adapted from Hecht, Norgaard, and Possio (n.d.).

To reach this conclusion, the authors developed a simulation model for a typical 20,000-hectare cattle ranch for which 75 percent of investment is provided by tax-credit funds. Internal rates of return (IRRs) to the investor's own resources (fresh money) and to all resources were then calculated under various assumptions regarding technology employed, intensity of grazing, and rates of land appreciation. The results under two different scenarios (table 7-3) show that livestock activities in Amazonia are profitable to corporations only when official subsidies or capital gains from land appreciation are present. The results also demonstrate that the IRR to a ranching corporation's own resources can be improved substantially through overgrazing. Indeed, cattle ranching can be made profitable only through overgrazing when cattle prices are low and when there is no land price appreciation. Overgrazing of course degrades the pasture and ultimately undermines the long-term viability of the ranch.

The findings of the simulation exercise described above have largely been borne out by field observations. On the environmental issue, Goodland (1980:18) rates cattle ranching as "the worst . . . of all conceivable alternatives" for Amazonia on the basis of its high potential for degrading the soil. Although some researchers debate this point, a comprehensive soil survey (involving eighty samples per age class of pasture) carried out in major cattle areas in eastern Amazonia lends it strong support (Hecht 1985). This survey indicates that the clearing of forest renders the potential of the soil low to marginal for pasture formation, particularly if physical changes such as soil compaction and the invasion of weeds are also considered. In practical terms, this means that stocking rates in Amazonia, which may be maintained at one animal per hectare during the initial years of pasture formation, typically decline to 0.25 animals per hectare after the fifth year.

Subsidized credit. Although livestock ranches benefiting from SUDAM-administered fiscal incentives have been major contributors to deforestation in southern Pará and northern Mato Grosso, they have not played a dominant role in this regard elsewhere in the region. Probably 90 percent of the pasture formation in Amazonia has been carried out by firms or individuals who have not received fiscal incentive funds. Clearly, other factors also played a role. One possibility, which is frequently mentioned in the literature, is the availability of subsidized rural credit (Browder 1988; Ledec 1985). Subsidized credit lines, like the fiscal incentive funds, increase private rates of return to investment and, as such, encourage activities—and by extension, deforestation—that would not be undertaken if credit were priced at market rates.

As shown in table 7-4, the volume of subsidized rural credit committed to Amazonia increased almost tenfold in real terms between 1974 and 1980.[4] In almost all years, the bulk of this credit was allocated to crop production, but the livestock sector also received large increases in the availability of subsidized credit after 1974. Much of this credit was extended through special lines with particularly attractive terms. For example, under the Program of Agricultural, Livestock, and Mineral Poles in Amazonia (POLAMAZONIA), a regional development program, a twenty-year investment credit was made available to ranchers at a nominal annual rate of 12 percent. The National Program of Livestock Development (PROPEC), which provided credit to ranchers in the more developed South and Southeast of Brazil at a nominal rate of 45 percent per year, offered terms similar to those of POLAMAZONIA to ranchers located in the Amazon region.

The effects of subsidized credit on the behavior of farmers and ranchers in Amazonia are difficult to quantify for several reasons. First, data at the farm and ranch level, such as size of operation, area cleared, output, and productivity, are extremely limited. Second, it is likely that a significant part of subsidized credit directed to agriculture and livestock was instead diverted to other uses. Although little is known about the extent of credit diversion in Amazonia, World Bank analysts estimated

Table 7-4. *Amazonia: Commitments of Official Rural Credit, 1970–85*

(millions of cruzados at 1985 prices)

Year	Total		Crops		Livestock	
	Cruzados	Percent	Cruzados	Percent	Cruzados	Percent
1970	106.0	100.0	78.2	73.8	27.8	26.2
1971	153.3	100.0	102.3	66.7	51.0	33.3
1972	264.2	100.0	157.2	59.5	107.0	40.5
1973	306.7	100.0	168.4	54.9	138.3	45.1
1974	203.9	100.0	118.7	58.2	85.2	41.8
1975	495.4	100.0	297.2	60.0	198.2	40.0
1976	899.5	100.0	417.4	46.4	482.1	53.6
1977	985.7	100.0	620.0	62.9	365.7	37.1
1978	1,332.0	100.0	901.8	67.7	430.2	32.3
1979	1,824.9	100.0	1,419.8	77.8	405.1	22.2
1980	1,882.6	100.0	1,724.5	91.6	158.1	8.4
1981	1,285.7	100.0	1,135.3	88.3	150.4	11.7
1982	870.5	100.0	637.2	73.2	233.3	26.8
1983	472.8	100.0	411.3	87.0	61.5	13.0
1984	198.2	100.0	175.4	88.5	22.8	11.5
1985	296.9	100.0	259.5	87.4	37.4	12.6

Source: IBGE (various years).

it to be on the order of 20 to 30 percent of the total rural credit in the late 1970s (Knight and others 1984:49). Finally, the influence of interest rate subsidies is inextricably mixed with that of the macroeconomic and sectoral policies they were supposed to offset.

Nevertheless, the availability of subsidized rural credit undoubtedly facilitated the acquisition and deforestation of large tracts of land in Amazonia, particularly during the latter half of the 1970s. (It is not possible, however, to specify how much less deforestation would have occurred in the absence of subsidized credit.) Moreover, because the special credit lines increased the unit subsidy element for undertakings in Amazonia more than in developed regions of Brazil, they probably attracted some resources that would have otherwise been invested in farms and ranches in the less fragile natural environments of the South and Southeast.

The National Integration Program

As shown in figure 7-2, increased road-building continued from the late 1960s into the 1970s. Considerable impetus to this construction was provided by the National Integration Program (PIN), established in

Figure 7-2. *Amazonian Road Network, 1960–85*

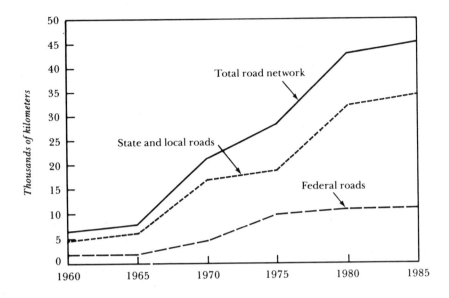

1970. PIN made resources available for some 15,000 kilometers of new road, including an east-west (Transamazon) highway connecting Amazonia with the Northeast, and a north-south (Cuiabá-Santarém) highway linking it with the South and Southeast. Plans were also made to build a second east-west highway (North Perimeter) along the northern bank of the Amazon River. A twenty-kilometer strip of land was to be reserved on either side of these highways for agricultural settlement projects.

The first concrete result of PIN was the completion of the initial 1,200-kilometer stretch of the Transamazon highway in late l972.[5] This part of the highway, which intersects with both the Belém-Brasília and Cuiabá-Santarém highways, was where the government hoped to relocate some of the Northeast's "excess" rural population. Government plans called for settling 70,000 families between 1972 and 1974 (Mahar 1979). To carry out the settlement program, the National Institute for Colonization and Agrarian Reform (INCRA)[6] established a network of new villages, towns, and cities at predetermined locations along the highway and demarcated 100-hectare farm lots nearby. The government actively recruited colonists in both the minifundia areas of the South and the latifundia areas of the Northeast through massive propaganda campaigns that promised attractive benefit packages.

Despite the huge amount of human and financial resources expended, the accomplishments of the PIN-financed road-building and directed settlement program were extremely modest. By the end of 1974, only about 5,700 families had been effectively settled along the Transamazon highway, less than 10 percent of the target set by the government (Moran 1981). By the end of the decade, this total had risen to only 8,000 families, of which just 40 percent had originated in the Northeast. Early in the settlement period, hopes that new natural resources would be uncovered during highway construction were shattered. No sizable mineral deposits were discovered; as of the early 1980s, only one company was reported to be extracting minerals (tin ore) along the Transamazon highway (Smith 1981).

The directed settlement schemes of the early and mid-1970s failed to create self-sustaining agricultural communities for many reasons. Certainly, adverse environmental factors played a major role. Both the routing of the Transamazon highway and the layout of the colonization projects were done hastily, with little regard for soil fertility or topography. As a result, cleared land eroded rapidly, which necessitated expensive maintenance work on the highway and the burning of additional forest in order to restore lost soil fertility. Institutional factors also played a role. Government planners and extension workers based agricultural development on annual crops, particularly upland rice, which are generally considered to be environmentally and economically unsustainable in areas cleared from tropical rainforests, except under very favorable circumstances (Fearnside 1983; Goodland 1980). By and large, the Transamazon settlers found themselves a long distance from markets, with inadequate rural infrastructure and services, and in areas with a high incidence of malaria.

The failure of the PIN-financed schemes to attract many migrants or to stimulate much economic activity helped to mitigate the potentially negative effects on the environment. Substantial deforestation did occur but it was concentrated along the Transamazon highway in the settlement areas of central Pará. Several other roads built during the 1970s—Cuiabá-Santarém, Porto Velho–Manaus, Northern Perimeter—also had limited environmental impacts. The major difference between these highways and those built in the 1960s (Belém-Brasília and Cuiabá–Porto Velho) is that the latter effectively linked the frontier with the country's urban-industrial centers while the former did not. According to one author, the ease of movement of commodities to and from these centers is "what counts most with regard to effects on settlement" and, by extension, on the environment (Sawyer 1984). The fact that no important mineral deposits or areas of fertile soils were discovered along the highways also protected the rainforest.

The Cuiabá–Porto Velho Highway and POLONOROESTE

In 1968, just a few years after the completion of the Belém-Brasília highway, the construction of another road—the 1,500-kilometer Cuiabá–Porto Velho highway (BR-364)—opened up the 243,000-square-kilometer state (then a federal territory) of Rondônia for settlement. This part of western Amazonia was a rich rubber-producing area at the turn of the century and in the mid-1950s became an important source of cassiterite (tin) and gold. Until the latter part of the 1960s, however, Rondônia, like most other parts of Amazonia, was virtually inaccessible to the rest of Brazil: to reach the southern part took several weeks by ship and boat along the Madeira and Amazon rivers. Rondônia's population (mainly itinerant rubber-tappers and prospectors) totaled only 70,000 in 1960, and practically all of the rainforest, which covered about 80 percent of the state, was still intact.

As in the case of the Belém-Brasília highway, the completion of BR-364 was followed by a wave of migrants, land-grabbers (*grileiros*), and adventurers seeking plots of fertile land that were purportedly free for the asking.[7] The migratory flow, which averaged perhaps 3,000 per year in the 1960s, increased tenfold in the ensuing decade (Mahar and others 1981). This increasing volume of migration was accompanied by major changes in the regional origins of the migrants. Whereas the vast majority of the original settlers in Rondônia were from the North and Northeast, most of the new contingent were experienced small-scale farmers from the southern state of Paraná. Large numbers also came from Mato Grosso, Minas Gerais, Espírito Santo, and São Paulo.

Both "pull" and "push" factors explain the sharp increase in migration to Rondônia after 1970. Two pull factors have predominated. First, the Cuiabá–Porto Velho highway route by chance traversed a few areas of relatively fertile soils, a fact that was publicized (and exaggerated) both by the government and by early settlers in their letters to family and friends. Second, official agricultural colonization projects allowed prospective settlers to obtain 100-hectare lots at a nominal price, along with basic services and infrastructure. The most important push factors are related to fundamental changes in the rural economy in Brazil's South and Southeast—the rapid spread of mechanized soybean and wheat production and cattle raising, killing frosts in coffee-growing areas, and the fragmentation of landholdings—that drastically reduced employment opportunities and caused widespread out-migration.

Rondônia's rapid population growth has had devastating effects on the rainforest. The speed at which this deforestation took place was, in some areas of the state, truly astonishing. The cleared area in the 80,000-hectare county (*município*) of Cacoal, for example, increased from 2,150 hectares in 1975 to 66,950 in 1978 (Fearnside 1982). Most of the deforestation was the result of clearing for agricultural purposes in official settlement projects along the main highway. Owing largely to the inadequacy of infrastructure, technical and financial assistance, agricultural research, and marketing facilities, however, most of the early settlers engaged in traditional, environmentally unsound farming practices (Mueller 1980). This usually involved the clearing and burning of a patch of forest and the cultivation of annual crops for one to three years, depending on soil fertility. Pasture would then be established on the original patch and the cycle would begin again with additional forest clearance.

By providing access to remote areas, Rondônia's network of feeder roads, which increased more than fivefold between 1975 and 1980 alone, greatly facilitated the deforestation process. Moreover, the poor condition of these roads and of BR-364 (especially during the rainy season) made it difficult to transport commodities to market and thus, along with other factors, discouraged the cultivation of tree crops such as cocoa, coffee, and rubber, which would have been much more appropriate from an environmental standpoint. Tree crops are strongly preferred over annuals and pasture because of their superior ability to protect the fragile soils of Amazonia from erosion. They also have some important socioeconomic advantages: their cultivation is labor-intensive and under reasonable market conditions can provide a decent standard of living for a farm family. The principal disadvantages of tree crops are their susceptibility to disease (such as the witches' broom fungus in cocoa and the leaf blight in rubber), their costly fertilizer requirements when grown on the poor land) typical of Amazonia, and a generally weak international market.

The growing socioeconomic problems in Rondônia led the government to reconstruct and pave BR-364 as part of a larger program of integrated regional development. Officially known as the Northwest Brazil Integrated Development Program (POLONOROESTE), the program covers a 410,000-square-kilometer area known as the Northwest, which includes all of Rondônia and part of western Mato Grosso. From the agricultural and environmental standpoints, the principal objective of POLONOROESTE was to reduce forest clearance on land without long-term productive potential and to promote, instead, sustainable farming systems based on tree crops. Land use surveys were conducted to identify areas with high potential, where new access roads, social infrastructure, agricultural research and extension, input supplies, crop storage, marketing, and farm credit could be concentrated (FAO/CP 1987). Environmental protection services in the Northwest were also to be strengthened.

The available data clearly suggest that the actions carried out under

POLONOROESTE so far have not succeeded in slowing down the pace of deforestation, nor have they appreciably altered traditional land-use patterns. In fact, the average area deforested annually in this decade has been more or less equal to the total deforested area in 1980 (see table 7-1). The expected major shift of farmland into tree crops has not materialized, and instead there has been a rapid conversion of forest into pasture (table 7-5). As discussed previously, pasture is one of the least desirable forms of land use in Amazonia from the environmental point of view.

A number of factors have contributed to the accelerated deforestation and inappropriate land use currently observed in Rondônia. A major proximate cause was the substantial jump in the migratory flow that followed the paving of BR-364 in 1984. About 160,000 migrants per year entered Rondônia in 1984–86 compared with an average of 65,000 per year in 1980–83. As a result, the state's population has grown at an average annual rate of almost 14 percent since 1980, pushing the state's total population to an estimated 1.2 million in 1987. This rapid population growth greatly increased the already high pressures on the forest. Population growth alone does not, however, explain the extremely rapid pace of deforestation nor farmers' preference for pasture formation over the cultivation of tree crops. One must therefore consider the role of certain institutional and policy factors.

Two institutional factors may be mentioned. First, the federal Institute of Forestry Development (IBDF) has not been able to enforce the 50

Table 7-5. *Agricultural Land Use in Rondônia, 1970–85*

(square kilometers)

Year	Crops Annual[a]	Perennial	Pasture	Forest[b]	Total[c]
1970	323.7	127.2	410.1	15,031.1	16,316.4
Percent	2.0	0.8	2.5	92.1	100.0
1975	1,503.9	457.6	1,645.2	26,681.4	30,820.5
Percent	4.9	1.5	5.3	86.6	100.0
1980	2,425.8	1,701.8	5,101.8	41,461.1	52,236.3
Percent	4.6	3.3	9.8	79.4	100.0
1985	3,153.3	2,238.0	15,611.5[d]	39,903.7[d]	60,906.6
Percent	5.2	3.7	25.6	65.5	100.0

a. Includes fallow land.
b. Includes natural pastures.
c. Area under farms at time of census; includes land unsuitable for agricultural use.
d. Estimated.
Source: IBGE (1987); Rondônia Secretariat of Planning; and author's estimates.

percent rule, which prohibits landowners in Amazonia from clearing more than half of their holdings. It is reported that some settlers in Rondônia have already cleared as much as 90 percent of their lots. In addition to its being unenforceable, some scientists argue that the 50 percent rule may intensify the damage to the environment it is designed to prevent (Goodland and Irwin 1975:30). Animals and plants need a minimum area for their survival. For insects this minimum may be measured in square meters; for larger mammals, such as the jaguar, 500,000 or more hectares may be necessary to support a genetically viable population. The fifty-hectare reserve in a typical colonist's lot, therefore, will not sustain anywhere near the level of biological diversity found in undisturbed rainforest. On-farm reserves may also harbor plant and insect pests that attack the surrounding agricultural areas.

Second, the intensification of smallholder agriculture envisaged under POLONOROESTE was predicated on the assumption that subsidized credit would be made available to finance purchases of modern inputs. The use of fertilizers and other inputs was particularly important for those farmers who settled by INCRA on poorer soils. Early in the implementation of the program, however, austerity measures resulted in a reduction in both the subsidy element and supply of credit (see table 7-4). Most farmers in the Northwest were therefore effectively denied credit. But even when credit was available, many farmers were reluctant to use it because they felt that the subsidy element was not high enough to offset the risks associated with the cultivation of tree crops (Wilson 1985). The problems caused by a lack of credit were further compounded by an extension service that continued to promote high-input farm models in the settlement areas (FAO/CP 1987).

In addition to institutional factors, certain land and tax policies have encouraged (or at least have not discouraged) unnecessary deforestation and inappropriate land use. Indeed, without a substantial modification of these policies, it is doubtful that land use patterns in the Northwest can be improved. One case in point is INCRA's policy of accepting deforestation as evidence of land improvement; that is, a migrant in either an official settlement project or in an invaded area can obtain rights of possession simply by clearing the forest.[8] Both good and poor lands are deforested indiscriminately in this manner. The geographical extent of these rights are determined by multiplying the cleared area by three, up to a maximum of 270 hectares. Once obtained, rights of possession can be sold either formally or informally depending on whether or not the migrant has occupied the land long enough to qualify for a definitive title. Although some migrants with a serious interest in developing sustainable agriculture have benefited from this policy, many others have used it as a means of acquiring land for speculative purposes.

Calculations made by the FAO/World Bank Cooperative Program

(FAO/CP 1987) show that it is possible for a speculator to net the equivalent US$9,000 if he clears fourteen hectares of forest, plants pasture and subsistence crops for two years, and then sells the rights of possession acquired by doing so. This constitutes a large sum of money in Rondônia, where daily farm wages average less than US$6. Additional calculations by the FAO/CP show that even bonafide farmers who have planted tree crops, but because of poverty cannot hold out until the trees are of bearing age, stand to make handsome profits by selling their lots after a few years (FAO/CP 1987:annex 5, table 1). Under the Brazilian income tax, such gains on land sales are theoretically taxed at a flat 25 percent rate. In practice, however, few of these capital gains are taxed, particularly in frontier areas, because land transactions are often informal and sales prices are underreported.[9]

The rural land tax (ITR), administered by INCRA, also encourages deforestation, at least in theory. The ITR was created in 1964 with the laudable objective of encouraging more productive use of land. The tax at present is assessed at a maximum rate of 3.5 percent of the market value of the land; the required 50 percent forest reserve is exempted from taxation. Reductions of up to 90 percent in the basic rate are given according to the degree of utilization of land (that is, the proportion cleared) and certain "efficiency" indicators established by INCRA, for example, crop yields, cattle stocking rates, and rubber extraction per hectare. In practice, the ITR probably has little influence on patterns of land use, mainly because the landowners declare the value of their land as well as the efficiency of its use. According to INCRA, only about half of all registered landowners in Rondônia paid any ITR in 1986; the average payment was the equivalent of only US$5 for those who did.

The Era of Big Projects

In the mid-1970s, the Brazilian government essentially abandoned the road-building and directed-settlement strategy embodied in PIN. In its place, the Program of Agricultural, Livestock, and Mineral Poles in Amazonia, or POLAMAZONIA, emphasized the development of large-scale, export-oriented projects in the livestock, forestry, and mining sectors in fifteen "growth poles" scattered throughout Amazonia. Before it was abolished in 1987, POLAMAZONIA was in essence a program of infrastructure development that, combined with existing fiscal and credit incentives, was aimed at creating a more favorable investment climate in Amazonia for private enterprise. The small-scale farmer, although relegated to a lower priority in regional plans, was not totally neglected. Recent official settlement programs, however, have been almost exclusively concerned with accommodating through POLONOROESTE the surge of spontaneous migration to Rondônia.

One factor explaining the abrupt shift away from directed small-scale settlement was general disillusionment with the Transamazon experience and the realization in government circles that Amazonia could not provide a quick fix to the demographic pressures and socioeconomic problems of the Northeast. Another important factor was the oil crisis of 1973, which hit Brazil, a major importer of this commodity, particularly hard. This external shock put into serious question the strategy of integrating Amazonia with the rest of Brazil on the basis of the automobile and truck. It also greatly increased the country's foreign exchange requirements to pay for oil imports and to service the rapidly mounting external debt. The government felt that Amazonian exports of minerals, timber, and agricultural products had the potential to make an important contribution to Brazil's annual foreign exchange earnings.

A key element of this new strategy was the exploitation of the immense natural resources of the Carajás subregion of eastern Amazonia. In order to pursue this objective in a rational manner, the government established the Greater Carajás Program (PGC) in 1980. The PGC is administered by an Interministerial Council headed by the minister of planning. The program area includes 895,000 square kilometers, or more than 10 percent of Brazil's total land area. Within this area, firms approved by the Interministerial Council enjoy generous fiscal incentives, government guarantees of foreign and domestic credit operations, and subsidized energy from the nearby Tucurui hydroelectric facility.

The Carajás Iron Ore Project, the first major project in the subregion, began in 1983 and is now fully operational. It was carried out under the responsibility of the Companhia Vale do Rio Doce (CVRD), a Brazilian parastatal mining enterprise. In contrast to most other projects in Amazonia, the Carajás Iron Ore Project was developed with close attention to its possible environmental impacts. Even before official approval of the Carajás project, CVRD commissioned a series of environmental baseline studies of the proposed project area, which covered climatology, ecology, botany, and related disciplines. On the basis of these studies the company established policies regarding environmental concerns such as forest clearing, topsoil stockpiling, erosion control, vegetation regeneration, and protection of fauna that eventually became environmental components of the project (Freitas 1982).

Between 1981 and 1985, CVRD spent around US$54 million on environmental activities related to the Carajás project (Kohlhepp 1987). These activities included land reclamation, the creation of protected natural reserves, and the promotion of environmental awareness and training. CVRD also maintained tight control over physical access to the project area to prevent unplanned human occupation. To oversee the implementation of the project's environmental components, CVRD created an independent group (GEAMAM), which consists of nine senior scientists who are to visit the project site for fifteen years. In addition, internal environmental commissions (CIMAs), made up of CVRD employees and contractors and coordinated by an ecologist, were placed on site permanently to make sure that operations adhere to government and company environmental guidelines (Freitas 1982).

In contrast to the orderly and environmentally responsible development of the Carajás iron ore reserves, the situation in the PGC area outside of the CVRD concession has been chaotic. Migration to this subregion has been intense in recent years. People are drawn by employment opportunities in the construction industry as well as by the prospects of striking it rich in newly discovered gold fields. This new wave of migration has arrived in a subregion of Amazonia that had already been subjected to settlement and large-scale cattle ranching following the completion of the Belém-Brasília highway. The recent spurt in population growth has exacerbated the environmental degradation and violent conflicts over land rights that have characterized much of the PGC area for many years (Branford and Glock 1985).

The government has so far failed to come up with a realistic, environmentally sound development plan for the overall PGC area. In 1983, the Ministry of Agriculture proposed dividing the subregion into seven agricultural poles, which would include 238,000 hectares of soybeans grown with the use of mechanized equipment, 12,600 hectares of sugarcane, and 417,000 hectares of cattle pasture. In addition, the plan called for 3.6 million hectares to be set aside for eucalyptus plantations to provide charcoal for metallurgical use. The agricultural development plan has been severely criticized for promoting land uses that are not likely to be sustainable (Hall 1987; Fearnside 1986a). Only the plans to develop a metallurgical sector along the railway corridor on the basis of locally produced iron ore and charcoal are currently moving ahead.

Fifteen pig-iron and manganese-iron projects had been approved for special fiscal incentives by late 1987; at least two are expected to begin operations in 1988.[10] An equal number of projects are presently being considered. The implementation of proposed projects would, at a minimum, add appreciably to pressures on the forest. Preliminary estimates suggest that at full operation the pig-iron plants would require 1.2 million tons of charcoal per year. To satisfy this demand would require the cutting of between 90,000 and 200,000 hectares per year of forest depending on tree stands, density of species used, and other factors. Since wood from the proposed eucalyptus plantations would not be available until the seventh year after planting, total deforestation attributable to the pig-iron plants would reach between 540,000 and 1.2 million hectares. The environmental implications of such large-scale deforestation are clearly negative.

No analysis of the true economic costs of producing pig-iron using charcoal has been carried out. In financial terms (based on gross revenues from pig-iron exports of US$100–US$110 per ton), it is estimated that any fuel that costs less than US$70 per ton of charcoal equivalent would render the plants profitable. Charcoal produced from the virgin forest, which presently sells for US$27 per ton in the region, clearly falls into this category. But market prices for charcoal reflect only the cutting and transportation costs for the wood used in its manufacture. If the full environmental costs of deforestation were to be included in the price of charcoal production, it is by no means clear whether the plants would remain viable. A key question, therefore, is whether charcoal can be produced at competitive prices from exotic trees.

Unfortunately, the cost of producing charcoal from plantation-grown eucalyptus on the scale envisioned and under the agronomic and climatic conditions prevailing in the PGC area is not known with certainty. The dismal historical experience with homogeneous tree plantations in Amazonia suggests that the cost will be high. Fearnside (1987) estimates that to supply the proposed pig-iron plants with charcoal derived entirely from plantation-grown eucalyptus would involve the planting of 2.6 million hectares of trees. This is thirty-five times the size of the next largest eucalyptus plantation in Amazonia, the 76,000 hectares planted on the Jari holdings in northeastern Pará. Despite massive injections of capital and intensive experimentation and research, the Jari project's tree plantations proved to be far more expensive and less productive than originally thought and have yet to turn a profit (Fearnside 1987; Kinkead 1981).

All in all, the experience accumulated so far during the era of big projects suggests that the general impact of government policies on the Amazon rainforest has been negative. The implementation of the Carajás Iron Ore Project, however, has shown that it is possible to exploit the region's resources in a manner that minimizes environmental damage. But it must be emphasized that the success of the iron ore project is largely attributable to the intrinsic nature of mining—usually a small area is involved and production is not dependent on environmental factors such as soils and climate—and to CVRD's environmentally sensitive approach and its ability to control events in its concession area. These are circumstances that will not be easy to replicate elsewhere in Amazonia. This point is vividly illustrated by recent and projected developments in the Greater Carajás area outside of the CVRD concession, in the areas of large-scale livestock development scattered throughout Amazonia, and in Rondônia.

Conclusions and Recommendations

Over the past twenty-five years the Brazilian government's policies to develop the Amazon region have rarely been designed and carried out with due regard to their environmental consequences. The felling of the rainforest, which began on a major scale during the 1970s, continues to take place at an accelerated pace in many parts of the region. The forests of Rondônia and parts of eastern Amazonia, in particular, are being cleared at explosive rates. Much of this deforestation has benefited neither the regional population nor Brazilian society as a whole, except perhaps in the very short term. Despite decades of intense development effort, Amazonia still accounts for only an insignificant 3 percent of the national income.[11]

Many of today's problems can be traced back to the decision made in the mid-1960s to provide overland access to Amazonia without sufficient knowledge of the region's natural resource base and of how to develop it in a sustainable manner. This initial error was compounded by subsequent decisions to provide generous incentives to investors willing to undertake environmentally questionable livestock projects and, more recently, smelting projects in the PGC area. Official settlement projects have also contributed to deforestation, although it would be wrong to place all of the blame for this on the settlers. Pushed by poverty and skewed land distributions in their regions of origin, the settlers have merely responded to government-created incentives in the form of access roads, titles to public lands, various public services, and, in the case of the Transamazon scheme, subsistence allowances.

There is no doubt that rapid deforestation will continue if present policies remain unaltered. In areas where overland access already exists, much damage has already been done. In such areas, the government should do what it can to promote the recuperation of degraded and abandoned lands and thus help to restore the forest's biological diversity. In cases where the land is presently occupied by small-scale agriculturalists, the best course of action would be to increase public support—both technical and financial—for those activities that can provide a decent living for a farm family and minimize additional environmental damage. Such activities might include cultivation of tree crops, the gathering of forest products, subsistence livestock (dairy cows, pigs, and chickens, for example), or some mixed production system. Because of the relatively high costs of production inherent to a remote frontier area and the uncertain market prospects for many of Amazonia's key exports, this approach would probably require some degree of subsidization on the part of the government. Such subsidies could be justified on both environmental and equity grounds.

A different policy should be developed for rainforest areas for which overland access does not yet exist. This policy would differ considerably from past policies, which have focused on opening up Amazonia indiscriminately for small and large-scale agricultural and livestock development. In effect, it would introduce an alternative development model based on the region's comparative advantage in forest-based economic activities. Under this new approach, the government would not construct any new roads or provide infrastructure or services (particularly land titles) in the region until detailed land-use surveys were carried out. Once the appropriate surveys were completed and the productive potential of the land known, physical access would be permitted only under special circumstances. (In this regard, it would be useful to improve facilities for water transport in these areas in order to reduce pressures for additional road construction.) Lands found to have limited agricultural potential— virtually all of the *terra firme* of Amazonia—would, under this policy, be held in perpetuity as forest reserves closed to all development or as sites for environmentally benign activities such as rubber tapping and gathering of Brazil nuts, tourism, or sustained-yield logging.

Recent events in Brazil suggest that a change in regional development policy along the lines suggested above may be in the offing. The government of Rondônia, for example, proposed in mid-1987 that the entire state be subject to agroecological zoning. The federal government is considering the possibility of extending this concept to all of Amazonia. These are definitely steps in the right direction. It should be kept in mind, however, that the government first proposed agroecological zoning for Amazonia in the late 1970s. To this end, a special commission composed of academics and government representatives was set up to draft suitable legislation. The original draft legislation included a commitment to preserve 150 million hectares of the region, of which 100 million hectares were to be rainforest. The preservationist tone of this document was reduced considerably in subsequent revisions, reportedly because of intense lobbying efforts on the part of timber and cattle companies and private colonization firms. No version of the legislation was ever approved by Congress.

The success or failure of the new attempts to apply agroecological zoning in Amazonia will depend largely on the technical quality of the proposals, the strength and depth of political support for the concept, and the existence of an overall policy framework consistent with rational land use. Although it is too early to pass judgment on the first two factors, it is clear from the analysis in this paper that the policy framework is still not in place. Some recommendations on how the policy framework could be improved follow.

First, the government should stop providing fiscal incentives for livestock projects in Amazonia. Disbursements to projects already under implementation could continue, but only in cases where SUDAM has confirmed by field visits that such projects are not located in rainforest areas. More than two decades of experience has shown that livestock projects have been responsible for much environmental damage and returned little in the way of production or employment. Livestock projects may also be criticized from the equity standpoint since most of the benefits from the fiscal subsidy have accrued to a small group of wealthy investors who have used these resources to appropriate large tracts of land on the agricultural frontier. Clearly, SUDAM-approved livestock projects have not succeeded in generating the social benefits necessary to justify the continuation of government subsidies.

The possibility of abolishing all regional fiscal incentives has been considered as part of Brazil's overall tax reform program. But owing to strong lobbying on the part of regional and extraregional special interest groups, there is apparently little likelihood that this will occur in the near future. The most powerful lobby group supporting the fiscal incentives for cattle ranching has traditionally been the Association of Amazonian Entrepreneurs (AEA) based in São Paulo (Pompermayer 1984). In addition, there is local political support from those who do not want to eliminate what is viewed as an important source of investment capital for the region.

Second, the government should declare a moratorium on disbursements of fiscal incentive funds for any projects in the PGC area—such as the proposed pig iron plants—that would use charcoal derived from the rainforest as the principal source of energy. Projects of this type have the potential to cause considerable deforestation in return for the production of relatively low-value products. Although these projects would in theory be obliged to replace the forest with tree plantations, previous attempts to establish large-scale tree plantations in Amazonia have never succeeded. Before a final decision is made on these projects, further research should be carried out on the true economic costs (including the environmental costs) of the projects and the possibilities of employing alternative energy sources such as hydroelectricity and natural gas.

Third, INCRA should modify the policy that recognizes deforestation as a form of land improvement and, as such, grounds for granting rights of possession. This policy has encouraged felling of forest in areas with little or no agricultural potential. It has also fueled land speculation. In the future, INCRA should not grant rights of possession or definitive titles to any lots on poor soils. In areas with poor soils but with potential for extractive activities, the granting of long-term concessions should be considered. INCRA has recently proposed a modification of its land-use policies along these lines. It proposes to provide twenty- to thirty-year concessions to individuals (largely rubber-tappers already in the region) or producer associations that undertake environmentally sound extractive activities in designated areas. This approach, which fits well with current proposals calling for agroecological zoning, should be encouraged.

Fourth, IBDF should consider abolishing the 50 percent rule. It has been shown to be unenforceable in a frontier region such as Amazonia and provides little, if any, protection to the environment. In place of the 50 percent rule, legislation should be passed that expressly permits the formation of contiguous blocks of reserves equal to 50 percent of the total area under agriculture in a given region rather than 50 percent of each farmer's lot. Such reserves would help to maintain biological diversity, benefit agriculture, and increase the number of migrants that could be settled on better soils in already-occupied areas. Block reserves have been established on an experimental basis in some of the newer settlement areas of Rondônia. Although some problems have been reported—for example, illegal invasions of the block reserves and disputes among settlers over their individual rights to use these reserves—the experiment should be closely monitored and evaluated for its replicability in other parts of Amazonia.

Fifth, the government should increase its efforts to improve the administration of taxes, which if duly collected could have beneficial effects on land use. A more effective administration of the 25 percent tax on capital gains from land appreciation, for example, could help dampen speculative pressures. The progressive rural land tax (ITR) also has the potential to improve land-use patterns by penalizing those who engage in environmentally unsound activities. The structure of the tax would need to be modified for Amazonia, however, to allow land left in virgin forest to be considered productive and thus to qualify for the lowest tax rate. Administration of this tax would also need to be vastly improved.

The above list of recommendations is not exhaustive. It does not, for example, include measures to improve employment opportunities in northeastern and southern Brazil and hence reduce pressures on the rural poor to migrate to the Amazon frontier. Undoubtedly much could be accomplished in this regard. The five policy reforms that have been suggested, however, combined with a well-thought-out and executed zoning plan for Amazonia, would reduce further economic losses and eliminate much unnecessary deforestation in the future.

Notes

1. Two geographical concepts of Amazonia are commonly used in Brazil. "Legal Amazonia," which is used for regional planning purposes, is made up of seven states and territories (Acre, Amapá, Amazonas, Mato Grosso, Pará, Rondônia, and Roraima) and parts of two others (Goiás and Maranhão). The

definition used for statistical purposes (the "North Region" or "classic Amazonia") is made up of six states and territories and is smaller by 1 million square kilometers. To the extent possible, the classic Amazonia concept will be used in this paper because it most closely corresponds to the area in rainforest.

2. The tax-credit mechanism underwent a further modification in 1974. Current legislation allows firms only a 25 percent credit against their income tax liability. It also calls for the establishment of the Amazon Investment Fund (FINAM), a type of mutual fund managed by BASA. Today, firms taking advantage of the tax credit initially receive shares in FINAM. The fund, in turn, acquires shares of stock in firms carrying out projects approved by SUDAM. Investors may hold or sell their shares in the fund or trade them for corporate stock held by FINAM. Investors with their own projects may directly acquire shares of their own stock. All corporate stock acquired from the FINAM portfolio is nonnegotiable for a period of four years. (For details, see BASA 1981.)

3. The vast majority of these were approved by SUDAM before the enactment of the 1979 rule prohibiting livestock projects in rainforest areas.

4. The sharp decline in the availability of official rural credit since 1980 is a reflection of government austerity policies that have affected agriculture in all regions, not just Amazonia

5. An additional 1,000-kilometer stretch of the Transamazon highway was inaugurated in early 1974. The 1,800-kilometer Cuiabá-Santarém highway was completed in late 1976. Construction of the 2,500-kilometer Northern Perimeter highway was essentially abandoned in the late 1970s for financial and technical reasons.

6. Although INCRA was the name of the institution durng the period covered by this chapter, its name was recently changed to the Ministry of Agrarian Reform and Development (MIRAD).

7. About 10 percent of Rondônia's soils are considered to be of good quality. This appears to be considerably better than the average for Amazonia as a whole (Mahar and others 1981:58).

8. Settlers wishing to engage in extractive activities such as rubber tapping or the gathering of Brazil nuts that do not disturb the forest are particularly disadvantaged by this policy.

9. Binswanger (1987) argues that other provisions of the income tax code, which exempt virtually all agricultural income from taxation, tend to increase the demand for land on the part of higher-income individuals. These provisions thus contribute to a more rapid conversion of forest to agricultural uses, land price appreciation, and greater concentration of landownership.

10. The government offers several types of fiscal incentives to approved firms. The most generous of these allows firms to take a tax credit equal to 50 percent of their corporate income tax liabilities on income earned within the PGC area for a period of ten years, if this money is reinvested in projects approved by the PGC Interministerial Council. (The tax credit was reduced to 50 percent from 100 percent in 1985.) "Fresh money" must account for at least 25 percent of any new investments using tax credit funds. To date, the beneficiaries of the tax credit provision have all been construction/engineering firms with profits from civil works activities in the PGC area. Other special fiscal in-

centives offered to firms located in the PGC area include exemptions from import duties and the federal excise tax (IPI).

11. This figure was calculated using standard national accounting procedures, which charge the depreciation of man-made assets such as buildings or equipment against current income, but not against the depletion of natural resources such as wildlife, minerals, or trees. If this anomaly were corrected, the real level of income generated in Amazonia would undoubtedly be much lower.

References

BASA (Banco da Amazonia). 1981. *FINAM: Legislação Básica dos Incentivos Fiscais para a Região Amazônica*. Belém: Banco da Amazônia.

Binswanger, Hans. 1987. "Fiscal and Legal Incentives with Environmental Effects on the Brazilian Amazon." Agriculture and Rural Development Department, World Bank, Washington, D.C. Processed.

Browder, John O. 1988. "Public Policy and Deforestation in the Brazilian Amazon." In *Public Policies and the Misuse of the World's Forest Resources*, edited by Robert Repetto and M. Gillis. Cambridge, U.K.: Cambridge University Press.

Branford, Sue, and Oriel Glock. 1985. *The Last Frontier: Fighting over Land in the Amazon*. London: Zed Books.

Denevan, William H. 1973. "Development and the Imminent Demise of the Amazon Rain Forest." *Professional Geographer* 25:130–35.

FAO/CP (Food and Agriculture Organization/World Bank Cooperative Program). 1987. "Brazil: Northwest I, II, and III Technical Review Final Report." Rome. Processed.

Fearnside, Philip M. 1982. "Deforestation in the Brazilian Amazon: How Fast Is It Occurring?" *Interciencia* 7(2):82–88.

———. 1983. "Development Alternatives in the Brazilian Amazon: An Ecological Evaluation." *Interciencia* 8(2):65–78.

———. 1986a. "Agricultural Plans for Brazil's Grande Carajás Program: Lost Opportunity for Sustainable Local Development?" *World Development* 14(3):385–409.

———. 1986b. *Human Carrying Capacity of the Brazilian Rainforest*. New York: Columbia University Press.

———. 1986c. "Spatial Concentration of Deforestation in the Brazilian Amazon." *Ambio* 15(2):74–81.

———. 1987. "Jari aos Dezoito Anos: Lições para os Planos Silviculturais em Carajás." In *Homem e Natureza na Amazônia*, edited by Gerd Kohlhepp and A. Schrader. Tübinger Geographische Studien 95. Tübingen University.

Freitas, Maria de Lourdes Davies de. 1982. "Brazil's Carajás Iron Ore Project: Environmental Aspects." Companhia Vale do Rio Doce, Rio de Janeiro. Processed.

Garcia Gasques, José, and C. Yokomizo. 1986. "Resultados de 20 Anos de

Incentivos Fiscais na Agropecuária da Amazônia." *XIV Encontro Nacional de Economia, ANPEC* 2:47–84.

Goodland, Robert J. A. 1980. "Environmental Ranking of Amazonian Development Projects in Brazil." *Environmental Conservation* 7(1):9–26.

Goodland, Robert J. A., and H. S. Irwin. 1975. *Amazon Jungle: Green Hell to Red Desert?* Amsterdam: Elsevier.

Guppy, Nicholas. 1984. "Tropical Deforestation: A Global View." *Foreign Affairs* 62(4):928–65.

Hall, Anthony. 1987. "Agrarian Crisis in Brazilian Amazonia: The Grande Carajás Programme." *Journal of Development Studies* 23(4):522–52.

Hecht, Susanna B. 1985. "Environment, Development and Politics: Capital Accumulation and the Livestock Sector in Eastern Amazonia." *World Development* 13(6):663–84.

Hecht, Susanna B., R. B. Norgaard, and G. Possio. n.d. "The Economics of Cattle Ranching in Eastern Amazonia." Graduate School of Architecture and Planning, University of California, Los Angeles. Processed.

IBGE (Instituto Brasileiro de Geografia e Estatística). 1983. *Censo Agropecuário—1980.* Rio de Janeiro.

_____. 1987. *Sinopse Preliminar do Censo Agropecuário—1985.* Rio de Janeiro.

_____. Various years. *Anuário Estatístico do Brasil.* Rio de Janeiro.

Kinkead, Gwen. 1981. "Trouble in D. K. Ludwig's Jungle." *Fortune* (August 20):102–17.

Knight, Peter, and others. 1984. *Brazil: Financial Systems Review.* Washington, D.C.: World Bank.

Kohlhepp, Gerd. 1987. "Problemas do Planejamento Regional e do Desenvolvimento Regional na Area do Programa Grande Carajás no Leste da Amazônia." In *Homen e Natureza na Amazônia,* edited by Gerd Kohlhepp and A. Schrader. Tübinger Geographische Studien 95. Tübingen University.

Ledec, George. 1985. "The Environmental Implications of Agricultural Credit Programs: Preliminary Findings for Brazil." Department of Political Science, University of California, Berkeley. Processed.

Mahar, Dennis J. 1979. *Frontier Development Policy in Brazil: A Study of Amazonia.* New York: Praeger Publishers.

Mahar, Dennis J., and others. 1981. *Brazil: Integrated Development of the Northwest Frontier.* Washington, D.C.: World Bank.

Moran, Emilio F. 1981. *Developing the Amazon.* Bloomington: Indiana University Press.

Mueller, Charles C. 1980. "Recent Frontier Expansion in Brazil: The Case of Rondônia." In *Land, People and Planning in Contemporary Amazonia,* edited by F. Barbira-Scazzocchio. Cambridge, U.K.: Cambridge University Press.

Myers, Norman. 1986. "Tropical Forests: Patterns of Depletion." In *Tropical Rain Forests and the World Atmosphere,* edited by G. T. Prance. Boulder, Colo.: Westview Press.

_____. 1984. *The Primary Source: Tropical Forests and Our Future.* New York: W. W. Norton.

Paula, Ruben D. de Garcia. 1971. "A Rodovia Belém-Brasília e os Fazedores de Desertos." *A Amazônia Brasileira em Foco* 6:78–95.

Pompermayer, Malori J. 1984. "Strategies of Private Capital in the Brazilian Amazon." In *Frontier Expansion in Amazonia,* edited by M. Schmink and C. H. Wood. Gainsville: University of Florida Press.

Santos, Roberto. 1980. *História Econômica da Amazônia: 1800–1920.* São Paulo: T. A. Queiroz.

Sawyer, Donald. 1984. "Frontier Expansion and Retraction in Brazil." In *Frontier Expansion in Amazonia,* edited by M. Schmink and C. H. Wood. Gainsville: University of Florida Press.

Smith, Nigel J. H. 1981. "Colonization Lessons from a Tropical Forest." *Science* 214(4522):755–61.

U.S. Department of Energy. 1986. *A Comparison of Tropical Forest Surveys.* Washington, D.C.

Wilson, John F. 1985. "Ariquemes: Settlement and Class in a Brazilian Frontier Town." Ph.D. diss., University of Florida.

Woodwell, George M., R. A. Houghton, and T. A. Stone. 1986. "Deforestation in the Brazilian Amazon Basin Measured by Satellite Imagery." In *Tropical Rain Forests and the World Atmosphere,* edited by G. T. Prance. Boulder, Colo.: Westview Press.

8

An Economic Justification for Rural Afforestation: The Case of Ethiopia

Kenneth J. Newcombe

The UNDP/World Bank assessment of the Ethiopian energy sector in 1983 recognized the large gap between supply and demand of firewood, the fuel of first choice for all rural and most urban households, and the need for a mammoth afforestation effort to reduce the imbalance.[1] A strategy was developed to remove the dependency of urban settlements on their rural hinterlands for fuel and to reestablish a dynamic equilibrium between the supply and demand for firewood in rural areas.

It is not difficult to establish the economic viability of peri-urban fuelwood plantations. Hard data exist on the present and long-run marginal costs of nonwood competitors, such as kerosene, liquefied petroleum gas (LPG), and electricity. These costs can be compared with the costs of establishing, maintaining, and harvesting fuelwood plantations. The only soft spot in such analysis is the derivation of an opportunity cost for the land resource so consumed. Establishing an equally rigorous economic justification for millions of hectares of rural afforestation is not so easy. For the great majority of rural settlements, the use of modern petroleum fuels is impractical—even impossible. Using petroleum as a proxy for an alternative fuel is in any case fiscally inconceivable; the foreign exchange will never be available for the million or so tons of kerosene per year that are implied. In practice, the available alternatives are already in use, and it is with the costs of their exploitation that the cost of rural afforestation must be compared.

The Conceptual Basis for Defining Costs

The basic questions are: If fuelwood stocks are depleted, what do rural Ethiopians use for cooking fuel, and what are the costs and benefits of these choices? Some answers become apparent by drawing a simple conceptual model of the process of adaptation and change that follows de-

Figure 8-1.   *Nutrient Cycles and the Deterioration of Ethiopian Agroecological Systems*

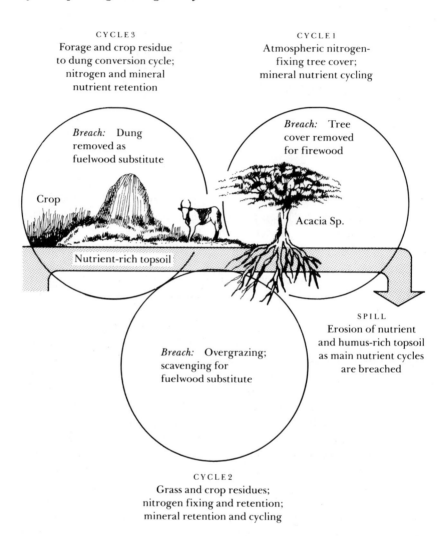

CYCLE 3
Forage and crop residue
to dung conversion cycle;
nitrogen and mineral
nutrient retention

CYCLE 1
Atmospheric nitrogen-
fixing tree cover;
mineral nutrient cycling

Breach: Dung
removed as
fuelwood substitute

Breach: Tree
cover removed
for firewood

Crop

Acacia Sp.

Nutrient-rich topsoil

SPILL
Erosion of nutrient
and humus-rich topsoil
as main nutrient cycles
are breached

Breach: Overgrazing;
scavenging for
fuelwood substitute

CYCLE 2
Grass and crop residues;
nitrogen fixing and retention;
mineral retention and cycling

forestation. The conceptual framework drawn here is based on limited fieldwork by others, limited observation and discussion by World Bank staff during the energy assessment, and basic principles of ecosystem behavior. Figure 8-1 illustrates three major nutrient cycles, which are sustained by natural energy flows that are operative to varying degrees in the smallholder cropping and grazing systems in Ethiopia. In ecological terms it can be argued that the system is not inherently stable; it is merely the rate of change that can be manipulated for economic reasons. In the simplest terms, there are five stages of ecological transition in Ethiopian smallholder systems.

Stage 1. The rate of timber harvested locally for all purposes (fuel, construction, tools, fences) exceeds, for the first time, the average rate of production. The existing timber resources are then progressively "mined"; firewood remains the main fuel source. Nutrient cycle 1 (figure 8-1) begins to decline, but the impact on food production is imperceptible. The general reason for the imbalance is population growth. The specific reasons include urbanization and major land clearing (for state farms, for example) whereby firewood and charcoal become cash crops. This leads to overcutting beyond local subsistence requirements.

Stage 2. Most of the timber produced on farms and surrounding land is sold elsewhere, to other rural and urban markets. Peasants begin to use cereal straw and dung for fuel: the relative proportions depend on the season. Nutrient cycles 2 and 3 are breached for the first time and nutrient cycling diminishes. Crop residues and dung are less available as inputs to the soil, and the lack of organic matter leads to poor soil structure, lower retention of available nutrients in the crop root zone, and reduced protection from the erosive effect of heavy rainfall. Hence, reserves of topsoil nutrients begin to decline (designated as "spill" in figure 8-1).

Stage 3. Almost all tree cover is removed. Now a high proportion of the cow dung produced and the woodier cereal crop stalks are systematically collected and stored, and both are sold for cash to urban markets. The fuel sold makes up a significant amount of the total cash income for the peasants. The yield of cereal crops and consequently the farm's ability to support livestock begin to decline. The number of draft animals and their power output are reduced, and the area being farmed also decreases. Soil erosion becomes serious. Nutrient cycle 1 ceases altogether.

Stage 4. Dung is the only source of fuel and has become a major source of cash. All dung that can be collected is collected. All crop residues are used for animal feed, though they are not sufficient for the purpose. Nutrient cycle 2 is negligible and cycle 3 is greatly reduced. Arable land and grazing land are bare most of the year. Soil erosion is dramatic and nutrient-rich topsoil is much depleted. The production of dung and dry matter has fallen to a small proportion of previous levels. The impact of dry periods is devastating because the buffering capacity of the ecosystem is negligible. Survival is marginal.

Stage 5. There is a total collapse in the production of organic matter, usually catalyzed by dry periods that previously were tolerable. Peasants abandon their land in search of food and other subsistence needs. Starvation is prevalent. Animal populations are devasted. Rural-to-urban migration swells city populations, increasing demand for food and fuel, and the impact of this urban demand is felt deeper into the hinterland (the urban shadow effect).

These five stages are part of a feedback loop with an increasing rate of change in the direction of stage 5, the terminal state. Evidence of this change is seen in the rapidly escalating real prices for biomass fuels both in the open market of major urban areas and at wholesale suppliers on the highways of the hinterland. For example, the market price of firewood in Addis Ababa has increased at the rate of 9 percent per year during the past decade (Unesco 1979). A critical transition that precedes stage 1 in this scheme is the transition from felling trees to clear land for growing food to felling trees primarily for fuel. Much of the Ethiopian landscape was forested in the late nineteenth century. Thus it is evident that this first critical transition has already established the preconditions for the final stages depicted here.

Quantifiable Elements

The precise amount of nitrogen fixation by leguminous trees and crops and by open fallow grassland is not precisely known for most ecosystems; certainly no data exist for Ethiopia. There is some information on nutrient loss through the erosion of topsoil for East Africa, but not for Ethiopia, and regional variability negates the option of using data from other areas. The loss of dung from nutrient cycles (cycle 2, figure 8-1) because of its use as a cooking fuel has been quantified to some extent, as has the smallholder cropping system in some typical central highland areas. Farmers in Ethiopia value dung and crop residues as fertilizers and soil conditioners. It is the rule, rather than the exception, for them to collect dung from corrals and nearby areas for storage and recycling as fertilizer and for use as a building material. Where wood fuels are abundant, the use of dung for fuel appears to be limited to specialized baking. Similarly, crop residues are quite purposefully retained not only for food, but also for fertilizer. It is only as the fuelwood supply declines markedly that these organic materials are commonly stored for later use or sold as fuel. In this case it may be that when fuelwood once again becomes available, dung and some crop residues will be collected, stored, and recycled as fertilizers and, furthermore, that these natural fertilizers will be directly applied to cultivated land.

Methodology

First, the annual production of dung is estimated for cattle only, because this is the most readily collected and commonly used dung. In areas that are at stages 3 and 4, however, donkey and horse manure is sometimes mixed with cow dung for fuel. The use of dung as fuel is estimated regionally from the fuelwood supply-demand tables produced in the forestry review of the UNDP/World Bank (1984) Ethiopian energy assessment, with allowance made for the use of crop residues as fuel. These estimates are supported by survey data from settlements and households on their fuel supply and consumption. They provide a crude global view of the major mineral nutrients (nitrogen and phosphorus) lost to agricultural production.

Next, the micro level of subsistence production is reviewed to establish a usable estimate of the likely application rates of dung per hectare, based on the average area cultivated per farm and the amount of dung produced and removed for fuel use. Then the crop yield response is estimated for this level of additional nutrient supply, using fertilizer response curves established in Ethiopia for double nutrient (nitrogen and phosphorus) applications for each common cereal crop in each soil type. In economic terms, this last step is peripheral wherever chemical fertilizer could reasonably be delivered, because the value of the dung nutrients to the economy is determined by the real cost of delivering to the farm an equivalent quantity of artificial (here imported) fertilizer. The amount of nutrient actually available to the plant, and hence the value per unit of nutrient supplied by natural and by artificial fertilizers, is a complex issue not addressed here. The two fertilizer types are treated as equal despite evidence suggesting that natural fertilizers are more readily absorbed by plants and hence have greater economic value in this physical context.

Thus, the only benefit quantified here is that derived from the phosphorus and nitrogen in cattle dung used as a fertilizer instead of a fuel. The value of this benefit is likely to be substantially less than the overall net benefit of afforestation. A partial listing of the prospective benefits of afforestation is provided in table 8-1.

Analysis

Dung Production and Consumption

In tables 8-2 and 8-3, the number of cattle and their annual dung production are estimated by region. The 1981–82 Ethiopian livestock census estimated the total cattle population at 24.6 million. Dung production, which is measured at the moisture content at which it is burned—15 percent (wet basis)—was estimated at roughly 23.2 million tonnes.

Table 8-4 is derived from the forestry review that indicates the level at which fuelwood demand is satisfied, the fuelwood deficit, and the

Table 8-1. *Potential Benefits of Rural Afforestation*

Direct benefit or use	Effect
Dung recycling	
Phosphorus[a]	Increases crop yield
Nitrogen[a]	Increases crop yield
Potassium and other	Increases crop yield
Major plant nutrients[b]	Increases crop yield
Micronutrients	Increases crop yield
Humus	Increases access of plant to available nutrients; reduces soil loss
Crop residues	
Soil protection	Reduces erosive impact of wind and rain; adds humus
Humus	As above
Nitrogen	Increases crop yield
Animal feed	Increases animal production and related benefits
Tree cover	
Fuelwood	Stabilizes household energy supply
Construction timber	Improved habitation and basic tool supply
Forage	Increases animal production
Nitrogen fixation	Increases crop yield
Preservation of soil moisture	Increases crop yield
Mineral nutrient cycling	Increases crop yield
Soil protection	Reduces wind and water erosion; sustains crop production

a. Evaluated in this chapter.
b. Calcium, magnesium, and manganese.

Table 8-2. *Average Liveweight of Ethiopian Cattle*

Age distribution (years)	Million head of cattle	Percent	Kilograms per head
Under 1	3.2	15	50
1–<2	2.1	10	125
2–<3	2.1	10	200
3–<4	2.0	9	250
4+	12.5	56	250
Total (weighted average)	21.9	100	202.5

Note: Figures exclude Eritrea and Tigrai.
Source: World Bank estimates and Ethiopia, Central Statistical Office.

implied dung and straw consumption by region. A set of rules to estimate the level of dung and cereal straw use has been compiled based on
review of the available literature and observations made during the assessment (see the appendix to this chapter). It is immediately apparent
that, by the logic of this method, more than 90 percent of the cattle
dung produced in Eritrea and more than 60 percent of that for Tigrai
and Gondar is used as fuel. The Eritrean estimate is an exaggeration of
cattle dung use, though it is likely that the collection level is this high and
includes equine dung. An Italian company (CESEN) is surveying the
inflow of fuels to ninety towns in Ethiopia as part of an energy sector aid
project; it found that for some towns in Eritrea and Tigrai up to 90 percent of total household cooking is done with dung transported from the
rural hinterland. In stage 4, cereal straw was said to be used only for ani-

Table 8-3. *Estimated Number of Cattle and Dung Production*
by Region, Ethiopia, 1981–82

| | Population (thousands) | | Cattle/ humans | Annual dung production[a] (thousands of tonnes) | |
Region	Cattle	Human		Fresh weight (13% dry matter)[b]	Dry weight (85% dry matter)
Arsi	1,644	1,207	1.4	10,134	1,550
Bale	509	922	0.6	3,138	480
Gamo/Gofa	602	1,051	0.6	3,711	508
Gojam	2,316	2,141	1.1	14,277	2,184
Gonder	1,524	2,156	0.7	9,394	1,437
Hararghe	1,249	3,284	0.4	7,699	1,178
Illubador	450	849	0.5	2,724	424
Kaffa	1,323	1,691	0.8	8,155	1,247
Shoa	6,419	6,792	0.9	39,569	6,052
Sidamo	1,863	2,744	0.7	11,484	1,756
Wallega	1,815	2,116	0.9	11,188	1,711
Wollo	2,181	2,743	0.8	13,444	2,056
Tigrai	1,735	2,270	0.8	10,695	1,636
Eritrea	989	2,586	0.4	6,096	932
Total	24,619	32,552		151,708	23,151

a. The conversion factor of cattle to dung is 202.5 kilograms average liveweight and
83.4 kilograms fresh weight of dung production per day per tonne of liveweight (fresh
weight is at 87 percent moisture content, wet basis).

b. Base numbers estimated in the 1980–81 census of livestock. The assumption is that
because of drought and starvation the livestock population in these regions did not grow
noticeably within the previous year.

Sources: Table 8-2; Fogg (1971); Natural Academy of Science (1977); and Ethiopia,
Central Statistical Office, "Agriculture Sample Survey, 1981/82, Preliminary Results of
Livestock and Poultry," Addis Ababa, October 1982.

Table 8-4. Estimated Fuelwood Deficits and the Consumption of Dung and Crop Residues by Region, Ethiopia, 1981–82

Region	Demand satisfaction (percent)	Fuelwood deficit (thousands of cubic meters)	Dung equivalent (thousands of tonnes)	Cow dung available (thousands of tonnes at 15% mcwb)	Dung burnt in households (thousands of tonnes at 15% mc)	Straw and stover burnt (thousands of tonnes at 15% mc)
Arsi	50	613	423	1,550	85	310
Bale	40	574	396	480	198	182
Eritrea	10	1,769	1,220	932	932	264
GamoGofa	50	547	377	568	75	277
Gojam	40	1,292	891	2,184	446	408
Gondar	20	1,738	1,199	1,437	1,079	328
Hararghe	40	1,974	1,361	1,178	681	624
Illubador	95	45	31	424	3	26
Kaffa	95	86	59	1,247	6	49
Shoa	30	3,720	2,566	6,052	1,283	1,177
Sidamo	70	897	619	1,756	124	454
Tigrai	10	2,065	1,424	1,636	1,282	130
Wollega	60	884	610	1,711	122	446
Wollo	10	2,504	1,727	2,056	1,554	461
Total					7,870	5,136

mc = moisture content; mcwb = moisture content, wet basis

Note: Equivalence is established using the calorific value of dry cow dung (15 percent moisture content, wet basis), 13.8 MJ (megajoules) per kilogram, and one cubic meter of wood (solid) at 25 percent moisture content, wet basis (as sold), 500 kilograms per cubic meter basic density, and 14.3 MJ per kilogram, thus 9,524 MJ per cubic meter. Thus one tonne of dung equals 1.45 cubic meters of wood. Cereal straw is fifteen MJ per kilogram, thus one tonne of straw equals 1.09 tonne of dung, or 1.57 cubic meters of wood. In all, the combustion efficiency of straw/stover and dung is assumed to be the same as for wood (that is, 8 percent with clay pots on open, three-stone fires).

The default allocation of dung and crop residues to make up the deficit in fuelwood supply is estimated according to the appendix to this chapter.

Source: World Bank estimates and UNDP/World Bank fieldwork for energy assessment, 1983.

mal feed; hence, there is a strong need to harvest all dung. This practice is facilitated by allowing children and shepherds to keep the money they make from the sale of dung collected from roaming cattle. The observation by staff of the International Livestock Center for Africa (ILCA) that some equine dung is mixed in with the cattle dung for sale in the more firewood-deficient areas lends weight to the conclusion that practically all cow dung is collected in the most deforested regions. The estimated amount of dung used for fuel is 7.9 million tonnes plus 5.1 million tonnes of crop residues.

Value of Dung

The composition of fresh and dry (85 percent dry matter) cow manure is shown in table 8-5. Elemental nitrogen is taken as 1.46 percent of oven-dry weight and elemental phosphorus is 1.30 percent of oven-dry weight. One tonne of urea contains the nitrogen equivalent of 32 tonnes of dry manure. The price for urea during the first quarter of 1983 was US$162.50 CIF (cost, insurance, and freight) per tonne bagged at Asseb. Shadow valued for foreign exchange, this becomes US$216.13 per tonne; and delivered at a rate of US$1.03 per vehicle kilometer at a conservative average of 800 kilometers inland, this becomes US$253.55 per tonne delivered, or US$0.54 per kilogram of nitrogen. Thus, each tonne of cow dung is worth US$7.88 for the nitrogen it contains. Diammonium phosphate (DAP) contains both nutrients in

Table 8-5. *Indices of Manure Production and Composition*

Daily production of manure per tonne of live animal (beef cattle)	83.4	kilograms
Moisture content		
Fresh	87	percent
Dry (as burnt)	15	percent
Composition (by weight)		
Nitrogen		
Wet	0.7	percent
Dry	1.46	percent
Phosphorus		
Wet	0.2	percent
Dry	1.30	percent

Note: For the most part, phosphorus is retained in the dry matter and is taken as conserved during drying. Nitrogen is readily volatilized (mainly the ammoniacal nitrogen) and is reduced almost fourfold during storage and processing. As stored, cow dung has been measured at 1.84 percent nitrogen. When spread on fields another 20 percent loss of nitrogen occurs.

Sources: National Academy of Sciences (1977); Porter and Young (1975); Newell (1980).

proportions roughly equivalent to dry cow dung. It is also the fertilizer most commonly applied in Ethiopia. Its average price in 1983 was US$233 CIF at Asseb. Its composition is 21 percent nitrogen and 23 percent phosphorus, or 44 percent of these combined, giving a price of US$0.53 per kilogram of nutrient (nitrogen and phosphorus) CIF, and US$0.79 per kilogram (nitrogen plus phosphorus) delivered 800 kilometers inland. One tonne of DAP roughly equals 16 tonnes of dry dung, giving a value for the latter of US$31.70 per tonne. This assessment of value obviously excludes various other nutrients contained in dung, which can contribute to plant growth. The composition of cow dung listed in table 8-6 can be expected to approximate that of Ethiopian cattle, though manure nutrients are somewhat site-related, that is, dependent on the composition of animal feed.

Potential for Increased Crop Yield

Although in theory the economic value of dung fertilizer can be equated directly to the delivered economic cost of imported equivalents, no account is taken of either the fiscal constraints of importing the quantity of fertilizer implied—490,000 tonnes of DAP costing US$114 million (CIF)—or of the minimal capacity to distribute this imported fertilizer. In reality, it is not possible to deliver fertilizer to more than 20 to 30 percent of the peasant farmers. Given these real constraints, the economic value of dung at the margin should be related to the market value of the incremental grain production arising from its application to the crop. This value is derived from quadratic response curves that have been fitted to data obtained from demonstration trials on farmers' fields conducted by the Ministry of Agriculture. These Ethiopian fertilizer response curves (NCC 1983) have been fitted separately for nitrogen and phosphorus, but the combined nutrient response curves are applicable here. Parabolic response surfaces were also fitted to 1980

Table 8-6. *Average Additional Major Nutrients and Trace Elements in Fresh Animal Manure*

Major nutrients	Kilograms per oven-dry tonne	Trace elements	Parts per million oven-dry matter
Potassium (K)	5.7	Boron	20.2
Calcium (Ca)	1.4	Manganese	201.1
Iron (Fe)	0.1	Cobalt	1.0
Sulphur (S)	1.0	Zinc	96.2
Magnesium (Mg)	1.1	Molybdenum	2.1

Source: Porter and Young (1975); reproduced with permission of Ann Arbor Science, an imprint of Butterworth Publishers.

and 1981 data to determine whether the effects of nitrogen and phosphorus were uniformly additive. These curves were derived for each grain growing region, for each grain crop, and for each soil type and pH level. To simplify, in this analysis teff (*Eragrostis teff,* a local grain staple) is the target crop for dung application. For red soils, the nitrogen (N) and phosphorus (P) interaction is significant for teff, thus the specific surface response curve is used:

$$Y = 7.1 + 0.038N + 0.086P - 0.0004N2 - 0.00038P2 - 0.00063NP$$

where *Y* is overall yield, 7.1 quintals per hectare is the constant (a quintal = 100 kilograms), and N and P the amount of elemental fertilizer in kilograms per hectare (elemental equivalent). Except for red soils, the general response curve is adequate:

$$Y = 755 \text{ (kg)} + 7.64X - 0.0219 X2$$

where X = combined N + P application in kilograms.

In order to use these response curves to determine the increased yield resulting from dung application, it is necessary to establish a reliable rate of application per location of cultivated cereal cropland in typical smallholder farms. The only hard, indicative data available are from detailed ILCA surveys of two moderately degraded agricultural ecosystems in Shoa Province, the Debre Berhan and Debre Zeit areas.

Micro Context: Dung Use by Smallholders

The ILCA has compiled extensive survey data on the pattern of subsistence agriculture in the regions of Debre Zeit and Debre Berhan and has gathered additional reference material in the process of its analysis of agricultural productivity in these areas.[2] These data have been reworked here to derive the average dung production of cattle, horses, and donkeys for the settlement concerned and to determine the end use of this dung in order to establish the quantities that might be recycled per unit area (see tables 8-7 and 8-8).

Fifty-five to 65 percent of the cow dung produced is used in households as a cooking fuel, which implies that 60 to 75 percent of the dung produced is collected, if 5 to 10 percent of the dung is used in house construction and maintenance. Each family uses 2.1 to 2.5 tonnes (85 percent dry matter) of dung per year for fuel. Taken together with the average area devoted to cereal crops of 1.65 hectare per family per year, this indicates an average potential recycling of between 1.29 tonnes of dung per hectare (18.83 kilograms of nitrogen and 16.77 kilograms of phosphorus per hectare) and 1.52 tonnes per hectare (22.12 kilograms of nitrogen and 19.7 kilograms of phosphorus per hectare). This is clearly an underestimate as the sale of dung cakes to town markets is a

Table 8-7. Farm-Based Dung Production: Indicative Data

Livestock (and average liveweight)	Debre Ziet				Debre Berhan			
	Average number	Liveweight (kilograms)	Dung per year (wet weight) AUᵃ	Tonnes	Average number	Liveweight (kilograms)	Dung per year (wet weight) AUᵃ	Tonnes
Oxen (275 kilograms)	1.86	512			1.02	281		
Cows (200 kilograms)	0.93	186			1.45	290		
Heifers (125 kilograms)	0.33	41			0.88	110		
Bulls (150 kilograms)	0.48	72			0.69	104		
Calves (50 kilograms)	0.64	32			0.98	49		
Subtotal		843	3.4	25.7		834	3.3	25.4
Horses (250 kilograms)	0.05	13			1.12	280		
Donkeys and mules (100 kilograms)	0.98	31			1.81	181		
Subtotal		44	0.2	0.9		461	1.8	9.4
Total		887	3.6	26.6		1,295	5.1	34.8

Note: The dung production rate for beef cattle is 41.7 kilograms wet weight per 500 kilograms of liveweight per day. For horses and donkeys the rate is 28.0 kilograms wet weight per 500 kilograms liveweight per day.

a. Animal unit = 250 kilograms.

Source: "Livestock Holdings per Farmer," ILCA survey of sixty farms in the Central Highlands, 1980.

Table 8-8. Nutrient Value of Dung and On-Farm Supply and Demand

	Debre Zeit				Debre Berhan			
Item	Dung per year (tonnes)	Nitrogen (kilograms)	Phosphorus (kilograms)	Moisture (percent)	Dung per year (tonnes)	Nitrogen (kilograms)	Phosphorus (kilograms)	Moisture (percent)
Nutrient value of dung								
Cow	25.7	180.0	51.4	87	25.4	177.7	50.8	87
Horse and donkey	0.9	7.7	1.2	87	9.4	81.1	12.2	87
Total	26.6	187.7	52.6		34.8	258.8	63.0	
Dry weight of dung (tonnes)		Debre Zeit				Debre Berhan		
Cow (85% dry matter)		3.90				3.88		
Horse and donkey (85% dry matter)		0.19				1.99		
Average household consumption								
Dry cakes								
Weekly (kilograms)		41				48		
Annual (tonnes)		2.13				2.50		
Annual on-farm supply of and demand for cow dung (tonnes)								
Supply		3.90				3.88		
Consumption		2.13				2.50		
Surplus		1.77				1.38		

Note: The surplus is not a true surplus because some dung is used for house lining, flooring, and other purposes. With an estimated minimal 250 kilograms dry weight per year for this purpose, the surplus is 1.52 tonnes in Debre Zeit and 1.13 tonnes in Debre Berhan. This suggests a collection efficiency of 60 to 75 percent, excluding dung being sold in the town. Donkeys bring about 80 tonnes of dung cakes to Debre Berhan for the Saturday morning market alone (1983 observation); householders report dung as a cash crop accounting for 10 to 30 percent of total annual income; and cow dung is supplemented with donkey dung to help meet demand.

Source: ILCA survey, 1980.

well-recorded source of cash income. In Debre Berhan, which has a population of 22,000, 80 tonnes of dung are carried into the town on approximately 2,300 donkeys every Saturday morning, which is market day.

Crop Yield Response

Using the fertilizer response curves, the incremental yields per hectare for teff on selected smallholder farms are 171 kilograms per hectare for red soils and 244.4 kilograms per hectare for other soils in Debre Zeit, and 192 kilograms per hectare for red soils and 281 kilograms per hectare for other soils in Debre Berhan. These incremental yields are low in comparison with those that would be achieved in the north central, north, and northwestern areas, where as much as 90 percent of the cattle dung and some donkey dung is collected (see table 8-4), and the prospective rate of dung application is higher. The dangers of generalizing and extrapolating here are clear, so only the lower estimates of incremental yield have been used in the ensuing economic analyses. The value of these yields is related directly to the economic value of the grain, which may or may not be reflected accurately in the market price. In consultation with the World Bank's East Africa regional agriculture staff, a 1983 value of 85 Ethiopian cents per kilogram of wheat was used, plus a premium of 20 percent to reflect the consumers' preference for teff. This provides a value of 1.02 birr per kilogram of teff for the incremental production specified (2.07 birr = US$1.00). Hence the implied value per tonne of dung is as shown in table 8-9.

Costs of Rural Afforestation

It is necessary to determine whether the benefits of retaining dung as a fertilizer and soil conditioner significantly outweigh the costs of agroforestry and rural woodlots to produce the firewood necessary to replace it as a fuel. Some preliminary estimates of the costs of fuelwood production in a rural context are provided in table 8-10, based on the

Table 8-9. *Values of Incremental Yield per Hectare and Implied Value of Dung*

Soil	Value of incremental yield per hectare (birr)		Implied value of dung (birr per tonne)	
	Debre Zeit	*Debre Berhan*	*Debre Zeit*	*Debre Berhan*
Red soils	174.42	195.84	135.21	128.84
Other soils	249.29	286.62	193.25	188.57

Note: 2.07 birr = US$1.00.

Table 8-10. *Simple Cost Estimate of Rural Afforestation by Peasant Association and Peasant Farmers in Ethiopia*

Cost	Birr per hectare
National and regional administration	
Transport	8
Offices and housing	4
Office expenditure (annual)	10
Staff	40
Subtotal	62
Peasant association	
Nursery for seedlings (includes seed, building, services)	90
Civil works (roads, bridges, drainage)	90
Tools and sundries	8
Subtotal	188
Labor [a]	
Planting	190
Harvesting	14
Weeding	6
Subtotal	210
Total	460

a. Average costs of one birr per person, per day of work, given some donated labor and contributions in kind.

forestry and fuelwood review of the Ethiopian energy assessment (UNDP/World Bank 1984). The analysis includes a wide range of environments, productivity, and potential costs. The institutional vehicle for large-scale afforestation in Ethiopia is also described in the Ethiopian energy assessment report. Because of the substantial cost of physical, administrative, and technical services, costs of production of wood fuel range from 4 to 6 birr per cubic meter (solid volume) or 6 to 9 birr per tonne of dung equivalent to a tonne of firewood (see table 8-10).

Evaluation

The comparative economic costs and benefits of dung used as a fuel can be summarized as follows:

	Birre per tonne
Prices of dung in the Gondar market, 1983	99–238
Value of dung (nitrogen and phosphorus content only) equated to farm-gate costs of imported fertilizer	45

Value of dung (nitrogen and phosphorus content only) equated to incremental production of teff when applied as fertilizer in typical peasant agricultural system   128–89

Value of dung as a fuel equated to the cost of producing firewood from afforestation in Ethiopia (see table 8-9)   6–9

This analysis indicates that sound logic is applied by farmers who sell their dung to urban markets as a source of cash. In the present marketplace, dung returns more when sold as fuel than it does when used as a fertilizer to produce additional grain. Although the data indicate an overlap in the level of returns from these respective uses of dung, its collection and sale as a fuel is a direct and certain means of generating income compared with the considerable uncertainties of achieving higher grain yields and of measuring the benefit once achieved. Moreover, artificially low grain prices in many regions provide a strong incentive for selling dung as a cash crop. Nevertheless, rural afforestation produces, even in inhospitable environments, a fuel of higher quality for less than one-quarter of the cost of the dung, based on the delivered costs of the equivalent imported fertilizer.

In economic terms, it would appear that in rural areas afforestation is justified in order to replace dung with wood as fuel. Dung could then be used as fertilizer, and the peasants would get greater value for the dung than when it is used as fuel. Using the lowest estimates of fuelwood yields (approximately 10 cubic meters per hectare, mean annual increment) and assigning to dung first its imported fertilizer equivalent value, then its low and high grain production equivalent value, the economic rates of return (ERRs) to rural afforestation are 35 percent, 59 percent, and 70 percent, respectively (table 8-11).[3] Even with a yield of 5 cubic meters per hectare mean annual increment and equating dung to imported fertilizer, the ERR is 23 percent. Dung now in use as a fuel has a fertilizer value of about US$123 million per year and, according to these crude though arguably conservative estimates, could add 1 million to 1.5 million tonnes of grain to the annual crop at a value of 850 to 1,275 million birr (based on 85 Ethiopian cents per kilogram).

Many additional benefits accrue from rural afforestation—both immediate (for example, the provision of other major nutrients, micro nutrients, and humus) and longer-term (soil conservation and climate control)—that have not been quantified here (see table 8-1). It is equally true, however, that peasant agricultural systems in Ethiopia and in any other large country are diverse, and the benefits assumed in recycling dung will be achieved in the short term only to the extent that the dung is already being collected and stored (and hence is available for application to cropland) as an intrinsic part of daily life. In some regions, this pattern of dung management may have to be the subject of an agricultural extension program.

Recent Emprical Findings

It has been five years since the fieldwork for the analysis in this chapter was done. On the basis of the original findings, the Ethiopian government, with World Bank and other outside assistance, has started a program to deal with the urgent issues of urban and rural fuel supplies. A number of empirical findings have emerged.

At the time of the original surveys, urban use of dung for fuel was generally limited to the smaller cities such as Debre Markos. At that time, there was little evidence of the use of dung for fuel in Addis Ababa except on special, festive occasions such as Easter, when baking increases and dung is preferred by some for that purpose. The surveys conducted by government authorities and the World Bank in May 1983 showed no evidence of a regular dung supply to the city. In contrast, a biomass supply and marketing study undertaken under the current World Bank–financed Ethiopia energy project shows that 7.5 percent of all fuel carriers—men, women, children, and donkeys—now carry dung into Addis Ababa, compared with the negligible dung trade five years earlier. This substantial growth in the Addis Ababa dung market can presumably be explained by its price and availability in relation to that of firewood and other increasingly scarce and expensive woody biomass fuels such as charcoal. The latter are clearly the preferred household fuels.

Another important finding of some recent investigations is the economic benefit of dispersed tree cover in Ethiopian croplands. Poschen (1986) found that with the use of traditional cropping methods in the Hararghe highlands, there was a 56 percent increase in yields of corn and sorghum planted under canopies of *Acacia albida*. Poschen's findings provide an additional indirect indication of the total costs of further reduction of tree cover in agricultural areas. Although these findings point to the desirability of having farmers plant their own trees, it is essential to prove to the farmers that the trees planted will enable them to reap a bigger harvest. The problem of land tenure in modern Ethiopia, including the villagization program, is a key constraint to such an option.

Conclusions

In Ethiopia, as elsewhere in the semi-arid sub-Sahelian zones of Africa, population pressure has led to extensive land clearing and the gradual

Table 8-11. *Economic Benefits of Rural Afforestation Derived from Substituting Firewood for Animal Dung as Fuel*
(per hectare)

	Cost stream		Benefits stream A		Benefits stream B		Benefits stream C	
	Rural woodlot establishment	*Yield of wood (tonnes, 25% mc)*	*Value of dung at imported fertilizer equivalent*	*Net cash flows (3)–(1)*	*Lower value of dung as worth of incremental grain production*	*Net cash flows (5)–(1)*	*Higher value of dung as worth of incremental grain production*	*Net cash flows (7)–(1)*
Year	*(1)*	*(2)*	*(3)*	*(4)*	*(5)*	*(6)*	*(7)*	*(8)*
1	171 [a]			−171		−171		−171
2	190 [b]			−190		−190		−190
3	3 [c]			−3		−3		−3
4								
5								
6								
7	5 [c]	33.5 [d]	1,507 [e]	1,502	4,288 [f]	4,283	6,331 [g]	6,326
8	2 [c]			−2		−2		
9								
10								
11								

12	5	33.5	1,507	1,502	4,288	4,283	6,331	6,326
13	2			−2		−2		
14								
15								
16								
17	5	33.5	1,507	1,502	4,288	4,283	6,331	6,326
18	2			−2		−2		
19								
20								
21	5	33.5	1,507	1,502	4,288	4,283	6,331	6,326
22	2			−2		−2		
Economic rate of return (percent)				35		59		70

mc = moisture content; mcwb = moisture content, wet basis

a. Consists of 31 birr per hectare as present value of overhead cash for administration, housing, and transport for a 200,000-hectare-per-year target program of afforestation; 90 birr per hectare for nurseries; 50 birr per hectare for civil works (see table 8-10).
b. 190 birr for planting at 1 birr per work day.
c. Harvesting and weeding costs.
d. One tonne of air-dry wood at 25 percent mcwb is roughly equal to one tonne of dry dung (15 percent mcwb)
e. 33.5 tonnes of dung × 45 birr per tonne.
f. 33.5 tonnes of dung × 128 birr per tonne.
g. 33.5 tonnes of dung × 189 birr per tonne.

removal of remaining trees for firewood. In the absence of other fuel, dung and crop residues are burned for domestic uses and sold to urban areas. But dung and crop residues have alternative value as fertilizer and soil conditioners. The analysis shows that retaining them in these uses and providing needed household fuels through systematic agroforestry programs is a highly attractive, long-term strategy for supplying household fuels and increasing agricultural output. Internal economic rates of return, based on the comparative evaluation of the costs of planting and growing new trees and the net productive increment in grain production from the retention of dung as a fertilizer, range between 35 and 70 percent in real terms. These are very high rates indeed. They are even more attractive because rural afforestation programs require very little foreign exchange for their implementation. Such programs, therefore, provide an attractive medium- to long-term rural development strategy with important benefits of added energy supplies, increased agricultural output, and environmental protection.

Appendix. Rules for Determining the Use of Dung as a Fuel

At least 5 percent of the dung produced is lost through rain each year. At least 5 percent additional is not collected for various reasons. Thus, the highest collection rate is 90 percent. Of this, at least 5 percent is used for building, except in an extreme shortage of fuel, when stones become the major building material.

In areas of extreme deforestation only 20 percent or less of the demand for firewood is satisfied. (The percentage of demand satisfaction is the percentage of the estimated demand for fuelwood that is met by the mean annual increment of forest and other tree cover production.) Dung is used almost exclusively as a fuel, and up to 90 percent of production is so allocated. Cereal straw is used almost entirely to feed animals and the remainder is used for building, for fertilizer, and for fuel. Where all the cow dung appears to be used, 100 percent of cow dung is attributed to consumption, the difference being made up by equine dung, which is used at the margin of cow dung supply. Thus, 90 percent of the fuelwood deficit is met by dung and 10 percent by crop residue.

In "moderate" areas where there is some tree cover and reasonable productivity (21 to 40 percent of demand for firewood is satisfied), dung is recycled partly as a fertilizer, and crop residue is used as a fuel as well as feed, with the residual again fertilizer. Thus, 50 percent of the fuelwood deficit is met by dung and 50 percent by crop residue.

In areas of fuelwood deficit but where significant tree cover remains (above 40 percent of the demand for fuelwood is satisfied), cereal straw is in excess of demand for fodder and local fuel and is traded as a surplus fuel to urban areas. Dung is mostly recycled as fertilizer and used as a building material, with minimal use as a fuel. Thus, 20 percent of the fuelwood deficit is met by dung and 80 percent by crop residue.

Notes

1. UNDP/World Bank (1984). The preference for wood as a fuel is a critical assumption from an analytical perspective. It was borne out by interviews with people in various consumer or decisionmaking positions, from peasants to department heads. First-hand observation and World Bank analysis of the household fuel cycle also confirm the preference for wood.

2. Personal communication from G. Gryseels, ILCA, May 1981.

3. The stream of costs are those required to establish a rural woodlot, using the generalized cost data provided in table 8-10. The corresponding stream of benefits is the production of wood in tonnes at 25 percent moisture content converted into the weight of dung with an equivalent lower heating value (where 1 tonne of dung equals 1.04 tonnes of firewood, approximately a 1:1 ratio) multiplied alternatively by the value per tonne of the dung in birr equated to the farm-gate costs of imported fertilizer, and to the incremental production of teff when dung is applied as a fertilizer in typical peasant agricultural systems.

References

Duckham, A. M., J. G. W. Jones, and E. G. Roberts, editors. 1976. *Food Production and Consumption: The Efficiency of Human Food Chains and Nutrient Cycles.* Amsterdam: North-Holland.

Fogg, E. C. 1971. "Livestock Waste Management and Conservation Plan." In *Proceedings of the International Symposium on Livestock Wastes.* Columbus: Ohio State University, April.

Hamilton, D. 1982. "The Status of Fertilizers in Ethiopia." Addis Ababa: National Chemical Corporation.

Hare, K. F. 1983. *Climate and Desertification: A Revised Analysis.* WEP-44. World Climate Application Programme. Geneva: World Meteorological Organization.

McCalla, T. M. 1975. "Characteristics, Processing and Use of Organic Materials." *FAO Soils Bulletin* 27:83–88.

National Academy of Sciences. 1977. "Methane Generation from Human, Animal and Agricultural Wastes." Washington, D.C.

National Chemical Corporation (NCC). 1983. "Results of Analysis of Fertilizer Response Data in Ethiopia: Supplements to the Status of Fertilizers in Ethiopia." Addis Ababa.

National Research Council (NRC). 1983. *Agroforestry in the West African Sahel.* Washington, D.C.: National Academy Press.

Newell, P. J. 1980. "The Use of High Rate Contact Reactors for Energy Production and Waste and Treatment from Livestock Units." Seminar on Energy Conservation and the Use of Solar and Other Renewable Energies in Agriculture, Horticulture and Fish Cultures, Polytechnic of Central London, September 15–19.

Porter, K. S., and R. J. Young. 1975. *Nitrogen and Phosphorus: Food Production, Waste, and the Environment.* Ann Arbor, Mich.: Ann Arbor Science, an imprint of Butterworth Publishers.

Poschen, Peter. 1986. "An Evaluation of the Acacia Albida-Based Agroforestry Practices in the Hararghe Highlands of Eastern Ethiopia." *Agroforestry Systems* 4:129–143.

Unesco. 1979. "Tropical Grazingland Ecosystems." *Natural Resources Research* 16. Paris.

UNDP (United Nations Development Programme)/World Bank. 1984. "Ethiopia: Issues and Options in the Energy Sector." Washington, D.C.

9

Managing the Supply of and Demand for Fuelwood in Africa

Jane Armitage and Gunter Schramm

Use of biomass fuels, mainly fuelwood or derived charcoal and, to a lesser extent, crop residues and dung, account for from 40 percent to over 90 percent of the total amount of energy consumed in the various countries of Sub-Sahara Africa (table 9-1). Rapid population growth, which accelerates land clearing for agricultural purposes and increases the consumption of wood fuels, is causing drastic reductions of forest cover, with subsequent deleterious effects on the environment through increased runoff, erosion, siltation, and flood damage.[1] In many regions the increasing scarcity of wood supplies, particularly in and around areas of concentrated consumption (cities and agroindustries such as tobacco curing and tea drying), has led to sharply increasing real prices, as well as actual shortages. In most regions, substitutes such as petroleum fuels, coal, natural gas, or electricity are not readily available because of a pervasive lack of foreign exchange to pay for additional imports, the lack of infrastructure to supply substitutes, the absence of suitable appliances for the use of other fuels, or the prohibitive costs of other fuels.

Population growth is the driving force behind increased demands for fuel in urban areas. In 1960, the urban population accounted for 11 percent of the total demand for fuel; by 1980, this percentage had grown to 21 percent, an average annual growth rate of 5.9 percent. Projections for the year 2000 call for an urban share of 37 percent, or 234 million out of 639 million people (World Bank 1981). Clearly, the pressures on limited wood resources are immense, and without the prompt adoption of appropriate, long-term strategies the exhaustion of regional resources may well become the rule, rather than the exception.

Outside of the limited areas protected as national parks or forest reserves, uncontrolled exploitation of remaining forest resources is commonplace in most countries. Most governments have no reliable data on

The authors want to thank Peter A. Dewees, Dennis Anderson, and Stein Hansen for their comments on earlier versions of this paper.

Table 9-1. Total Energy Consumption by Source of Fuel

Country	Percentage of population in urban areas (1980)	Percentage of total energy use							Total (thousands of TOE)[b]
		Fuelwood	Charcoal	Other traditional fuels[a]	Electricity	Petroleum	Natural gas	Coal	
Low income									
Benin	14	78	1	8	1	11	n.a.	n.a.	847
Burkina Faso		87	1	5	1	6	n.a.	n.a.	1,787
Burundi	2	46	2	n.a.	13	38	n.a.	1	906
Ethiopia	15	37	1	55	1	7	n.a.	n.a.	8,016
Ghana	36	59	10	5	3	23	n.a.	n.a.	3,068
Guinea	18	85	2	n.a.	1	12	n.a.	n.a.	2,435
Kenya	14	65	9	n.a.	2	24	n.a.	1	8,070
Malawi	10	86	5	3	1	4	n.a.	1	3,321
Niger	13	86	n.a.	n.a.	2	12	n.a.	n.a.	964
Rwanda	4	76	1	17[c]	2	4	n.a.	n.a.	1,008
Sierra Leone	25	81	4	n.a.	2	13	n.a.	n.a.	911
Somalia	30	77	5	5	1	12	n.a.	n.a.	1,123
Sudan	25	45	29	9	1	17	n.a.	n.a.	6,148

Tanzania	12	88	4	n.a.	1	7	n.a.	n.a.	9,025
Togo	20	60	4	7	4	26	n.a.	n.a.	793
Uganda	12	90	4	n.a.	1	6	n.a.	n.a.	4,513
Zaire	34	75	5	7	4	7	n.a.	2	8,598
Middle income									
Botswana	n.a.	48	n.a.	n.a.	16	20	n.a.	15	725
Congo, P.R.	45	44	1	1	5	49	n.a.	n.a.	574
Côte d'Ivoire	38	52	4	9	5	31	n.a.	n.a.	2,765
Liberia	33	61	8	n.a.	9	23	n.a.	n.a.	917
Mauritania	23	44	6	3	4	43	n.a.	n.a.	289
Nigeria	20	60	1	n.a.	2	32	5	1	22,016
Senegal	25	57	7	n.a.	3	33	n.a.	n.a.	1,679
Zambia	38	25	6	1	39	18	n.a.	12	3,604
Zimbabwe	38	25	n.a.	3	29	11	n.a.	32	5,874

n.a. Not available.
a. Agricultural and animal residues.
b. Tons of oil equivalent.
c. Includes peat.
Source: UNDP/World Bank data.

their forest resources, have no control over them, and do not collect significant revenues from them. This is in spite of the fact that sales of firewood and charcoal for urban and industrial use represent a multimillion dollar business that employs tens of thousands of people.[2] One reason for the uncontrolled exploitation of the forest cover is the economic value of the standing stock of wood cannot be easily recovered from those who cut it down. In many countries, neither the land on which the wood is standing nor the wood itself belongs to those who cut it. Therefore, there is no incentive for long-term management that would lead to sustainable production of wood, nor is there an incentive to maximize yields. Even though wood cutting and charcoal making are often illegal, such activities are practically never policed. There is no incentive for replenishing the dwindling supplies of wood through private replanting, because prevailing market prices reflect only the costs of production from standing stocks, but not the additional high costs of replacement.

There are several reasons why market prices do not reflect the growing scarcity. The first is related to the stock-flow problem. Although in the past, in regions with little or no net population growth, wood extraction was probably roughly balanced by regrowth within economic hauling distance to centers of demand, this is no longer true. When cutting starts to exceed regrowth, stocks necessarily must decline. When the ratio of standing stocks to net incremental growth is in the neighborhood of forty to seventy, this decline is slow and imperceptible at first. Catastrophe approaches incrementally, much like a malignant cancer that may take decades to develop, but then ravages its victim within weeks or months. Second, competition in the commercial firewood and charcoal business, combined with the fact that the resource base is common property, keeps prices low, reflecting only cutting, production, and distribution costs until the local resource base essentially has disappeared.

Third, in many countries the wood for charcoal making comes mainly from agricultural land clearing operations.[3] Although this wood has no alternative value and would otherwise be burned as waste, it represents a nonrenewable resource. When land clearing operations come to an end, as ultimately they must when land suitable for agriculture runs out, supplies start to dry up, shortages occur, and prices rise drastically. Such price increases induce more poaching of remaining wood resources in protected, environmentally fragile areas (such as forest belts protecting important watersheds). They also cause overcutting of standing private tree stocks and single trees, which leads to increased wind and water erosion, as has happened in Ethiopia and Nigeria (see chapters 8 and 1 0).

The Rationale for the Continued Use of Wood Fuel

With the growing scarcity of wood fuels and the resulting environmental damages, the question often asked is "Why don't these countries follow the example of the rest of the world and convert rapidly to the use of so-called commercial energy sources?"[4] The answer to this question is quite complex and has both macro- and microeconomic dimensions.

Few of the Sub-Saharan countries own readily usable domestic resources of commercial fuels such as petroleum, natural gas, coal, or electricity. Only Angola, Cameroon, Gabon, Nigeria, People's Republic of the Congo, and Zaire have domestic crude oil resources, and some of them must export crude oil and import refined products for all of their domestic requirements because they lack suitable refineries.[5]

A few countries have coal (Botswana, Mozambique, Swaziland, Tanzania, Zambia, and Zimbabwe), but deposits often are far from centers of demand (for example, in Tanzania), and appliances for their use cost far too much for households, given the low prevailing income levels. This is true for commercial uses as well. Tobacco curing, for example, is a prodigious consumer of wood fuels.[6] But to replace wood with coal (as Zimbabwe did when wood fuels became scarce) requires more sophisticated and costlier furnaces as well as electricity to drive the required fans. In most of the tobacco-growing regions, electrici,ty is not available, however, and the costs of making it available are usually prohibitive because of the small loads involved.

Electricity and natural gas (in the few countries that own gas deposits) are other options, but are usually very expensive relative to other energy resources. In most countries, less than 20 percent of urban households have access to electricity, and fewer still can afford the high costs of electric energy for cooking.[7] Natural gas is even more costly than electricity. It has long been established that service connections to households are too expensive without winter heating or summer air-conditioning loads.[8] It is not surprising, therefore, that domestic distribution systems do not exist in any of the gas-producing countries in Sub-Sahara Africa. In developing countries where distribution systems do exist, domestic gas users are usually heavily subsidized (UNDP/World Bank data).

Kerosene and to a much lesser degree liquefied petroleum gas (LPG) are left as the only practical fuels that could be readily substituted for wood.[9] But in all of the Sub-Saharan countries, including oil-rich Nigeria, additional supplies of kerosene would have to be imported.[10] This would put a substantial strain on the balance of payment position of most African countries. As illustrated in table 9-2, replacement of all urban wood fuel consumption by kerosene would increase the overall demands for petroleum imports from a relatively modest 5 percent in Nigeria to as much as 74 percent in Tanzania.[11] For most Sub-Saharan countries, the impact of such an increase on foreign exchange balances would be serious. Petroleum imports already accounted for between 12 and 53 percent of total export earnings of these countries in recent years (see table 9-3).[12] Almost all of the Sub-Saharan countries incur

Table 9-2. *Impact of Using Kerosene Instead of Urban Wood Fuels*

Country	Year	Thousands of tonnes kerosene required to replace:			Percentage increase in total petroleum demand if kerosene substituted for:	
		Urban charcoal[a]	Urban firewood[b]	All urban wood fuels	Urban charcoal	All urban wood fuels
Low income						
Benin	1983	5	35	40	4	34
Burkina Faso	1983	3	n.a.	n.a.	n.a.	n.a.
Burundi	1980	5	10	15	15	46
Ethiopia	1982	45	125	170	9	32
Ghana	1985	130	50	180	19	26
Guinea	1984	20	90	110	5	29
Kenya	1985	180	155	335	11	20
Niger	1980	n.a.	35	35	n.a.	23
Nigeria	1980	50	355	405	1	5
Sierra Leone	1986	20	40[c]	60[c]	11	34[c]
Somalia	1984	25	50[c]	75[c]	14	41[c]
Sudan	1980	310	125	435	32	45
Tanzania	1981	150	320[d]	470[d]	24	74
Togo	1981	15	25	40	6	17
Uganda	1982	60	20	80	38	51
Zaire	1983	170	370	540	29	91
Middle income						
Côte d'Ivoire	1982	45	90	135	5	14
Liberia	1983	25	25	50	6	12
Mauritania	1984	5	8[c]	13[c]	3	8[c]
Senegal	1981	40	15	55	6	8
Zambia	1981	90	20	110	13	16
Zimbabwe	1980	5	60	65	1	11

n.a. Not available.

a. Charcoal stove efficiency is assumed to be 15 percent and kerosene stove efficiency 35 percent.

b. Firewood stove efficiency is assumed to be 10 percent and kerosene stove efficiency 35 percent.

c. It is assumed that urban use of fuel wood is 20 percent of the total demand for fuel wood in the absence of data on this proportion.

d. Urban use of fuel wood is assumed to be 15 percent of total demand in the absence of definitive data.

Source: UNDP/World Bank data.

substantial yearly foreign exchange deficits, and any increase in debt from any source could only worsen their already precarious economic conditions.[13]

Given the heavy burden of oil imports, it is not surprising that in many of the countries petroleum products, kerosene in particular, are commonly in short supply. Countries with their own refineries (such as Kenya, Nigeria, Tanzania, and Zaire) are usually geared to supply the major white products, diesel fuel and gasoline, but do not produce enough kerosene and its twin, jet fuel, to serve local demands. Additional net imports would be needed. Lacking foreign exchange, such imports are not forthcoming. Furthermore, in the majority of countries, posted kerosene prices are kept artificially low by the government on the mistaken belief that kerosene is a "poor man's fuel." This makes distribution margins for petroleum distributors unattractive, so that they make no special efforts to secure additional supplies even in those countries where additional imports could be secured and paid for. The result is shortages, supply interruptions, and black market operations.

Insecurity of supplies and high prices, in turn, make it unattractive for households to convert from wood fuels and charcoal to kerosene, even if, on the basis of posted prices, kerosene would appear to be the lower-priced fuel (see table 9-4).[14] After taking account of the average efficiencies of kerosene versus charcoal stoves, the ratio of kerosene to charcoal prices was less than unity in eleven out of the eighteen countries shown in table 9-4, indicating that kerosene was lower-priced than charcoal. It is important to note, however, that these comparisons are based on official prices only. In Sierra Leone and Tanzania, for which black market prices are also shown in the table, the ratios are substantially higher. Furthermore, in most of the countries with apparently low kerosene to charcoal price ratios, currencies were grossly overvalued in the years shown. Since 1986, devaluations of several hundred percent have taken place in Tanzania, Uganda, Zaire, Zambia, and several other countries. This means that the relative cost of imported kerosene has increased accordingly. In any case, proper shadowpricing of the overvalued exchange rates of most of the countries would result in considerably higher kerosene prices.

Compared with imported fuels, domestic wood fuels have several distinct advantages. As a domestic resource, they have only a modest impact on scarce foreign exchange, limited largely to the costs of transport—imported petroleum fuels, trucks, and spare parts. In Kenya, with average hauling distances of 200 kilometers, transport costs accounted for about 15 percent of the retail price of charcoal in Nairobi. Import costs of fuel, vehicles, and spare parts accounted for about 55 percent of total transport costs. Thus imports accounted for only about 8 percent of the delivered costs of charcoal (UNDP/World Bank data).

The second advantage is the creation of employment. Fuelwood cutting and charcoal production are rather labor-intensive activities. Even more important, they utilize unskilled or semiskilled labor, largely on a seasonal basis outside of peak agricultural activity periods. This type of employment is urgently needed in developing countries with a huge

Table 9-3. Petroleum Demand and Import Costs in Africa

(thousands of tonnes)

Country	Year	LPG	Gasoline	Kerosene Tonnes	Kerosene Percent[a]	Jet fuel	Diesel oil	Total petroleum product demand[b]	Net import petroleum cost (millions of current U.S. dollars)	Ratio of net petroleum import cost to total earnings from export
Low income										
Benin	1984	0.5	42.1	14.2	12	15.3	37.0	119.3	48.0	20
Burkina Faso	1983	0.6	46.1	11.0	12	1.4	28.4	90.3	n.a.	n.a.
Burundi	1980	...	16.6	0.8	3	n.a.	15.2	32.6	n.a.	n.a.
Ethiopia	1982	4.7	117.6	10.6	2	75.4	240.0	526.0	185.0	53
Ghana[c]	1985	4.1	220.8	113.9	16	24.4	308.9	699.7	166.0	26
Guinea	1984	n.a.	70.4	3.1	1	9.0	99.1	382.9	n.a.	n.a.
Kenya	1980	21.5	310.8	84.1	5	347.9	451.1	1,677.5	297.5	36
Malawi	1980	...	39.5	5.7	4	11.7	77.3	138.2	56.0	22
Niger	1981	0.4	37.9	4.1	3	16.2	89.8	151.2	34.0	26
Rwanda	1980	...	24.3	6.7	14	n.a.	13.5	48.7	n.a.	n.a.
Sierra Leone	1984	0.9	38.2	27.7	16	11.4	75.7	178.5	27.0	24
Somalia	1984	0.1	42.0	13.6	7	1.6	99.4	181.6	54.7	40

Country	Year									
Sudan	1981	5.4	240.4	16.5	2	43.1	529.0	967.1	n.a.	n.a.
Tanzania	1982	5.4	122.0	68.9	11	37.7	294.9	638.0	289.0	n.a.
Togo	1982	0.6					41.7	239.4	58.0	22
Uganda	1982	0.3	43.9	28.9	18	17.3	50.3	156.7	100.0	30
Zaire	1983	0.3	89.4	38.7	7	98.9	296.6	591.6	n.a.	n.a.
Middle income										
Botswana	1982	1.1	47.6	4.2	3	0.8	87.0	141.1	91.0	22
Congo, P.C.	1985	3.9	53.3	21.1	8	10.7	122.1	250.5	n.a.	n.a.
Côte d'Ivoire	1983	16.4	203.7	63.3	7	60.0	280.4	958.5	n.a.	n.a.
Liberia	1983	0.5	66.3	5.0	1	27.9	123.3	402.1	115.0	24
Mauritania	1983	2.2	27.2	1.4	1	12.1	124.3	167.2	44.0	12
Nigeria	1981	48.0	3,620.0	900.0	11	460.0	2,280.0	8,248.0	n.a.	n.a.
Senegal	1981	10.4	102.9	10.7	2	141.3	144.1	704.9	144.0	51
Zambia	1981	1.0	120.0	30.0	4	65.0	287.0	683.0	240.0	19
Zimbabwe	1980	5.3	192.8	20.4	4	59.8	305.2	583.5	265.0	23

n.a. Not available.

. . . Negligible.

a. The percentage of kerosene use relative to total petroleum demand.

b. Total petroleum demand includes demand both for products shown and for residual fuel oil (not shown).

c. Sixty-one percent of the kerosene is used in urban households, 31 percent in rural households, and the remainder in industry and commerce.

Source: UNDP/World Bank data.

Table 9-4. *Retail Prices for Kerosene and Charcoal in Major Urban Centers*
(U.S. dollars per million BTU)

Country	City	Year	Kerosene	Charcoal	Unadjusted	Ratio of retail price of kerosene to charcoal adjusted[a]
Benin	Cotonou	1983	8.3	4.9	1.69	0.73
Burkina Faso	Ouagadougou	1984	12.0	4.8	2.50	1.07
Congo, P.R.	Brazzaville	1986	14.9	16.7	0.89	0.38
Côte d'Ivoire	Abdijan	1984	12.5	6.3	1.98	0.85
Ethiopia	Addis	1983	12.6[b]	13.6[b]	0.93	0.40
Ghana	Accra	1986	3.6[c]	6.90–0.52	0.22	—
Kenya	Nairobi	1985	12.2	4.3	2.84	1.22
Liberia	Monrovia	1984	19.0	5.2–8.1	3.02	1.01–1.57
Mauritania[d]	Nouakchott	1983	16.5[c]	5.0[c]	3.30	1.41
Niger	Niamey	1983	12.1	5.4	2.24	0.96
Senegal	Dakar	1983	12.3	5.5	2.24	0.96

Sierra Leone	Freetown	1986	5.3c	7.40	0.72	0.31
Somalia	Mogadishu	1985	15.1	4.90	3.08	1.32
Tanzania	Dar-es-Salaam	1983	7.9c	7.90c	1.00	0.43c
			16.5e	2.09e	—	0.90e
Togo	Lome	1983	11.2	3.80–5.10	2.20–2.95	0.94–1.26
Uganda	Kampala	1982	12.2f	3.9	3.13	1.34
Zaire	Kinshasa	1984	14.9	9.1	1.64	0.70
Zambia	Lusaka	1982	14.2	8.80	1.61	0.69

Note: Prices are converted to U.S. dollars per million BTU at official rates of exchange, assuming that 1 metric ton charcoal = 27.4 million BTU, and 1 barrel of kerosene = 5.2 million BTU.

a. This is the ratio of prices after accounting for relative differences in efficiencies of use. Assumed efficiencies are 35 percent for kerosene and 15 percent for charcoal stoves. A value equal to 1 implies equality of kerosene and charcoal prices.

b. Based on an estimated shadow exchange rate of 2.7 birr = US$1.00 compared with the official rate of 2.07 birr = US$1.00 in 1983.

c. Official controlled price.

d. In Mauritania about two-thirds of the charcoal supply is met by imports from Senegal.

e. Black market sales.

f. Exchange rate of 200 USh = US$1.00 (at the end of 1982).

Source: UNDP/World Bank data.

surplus of unskilled, unemployed labor. Properly shadow priced, the real economic costs of this employment are far lower than the monetary earnings of the people employed.

Offsetting these advantages are the environmental costs of continued overcutting of existing wood resources. The ultimate consequences have been graphically described in chapter 8, and do not have to be repeated here. The question is: Does continued use and cutting in excess of regrowth necessarily have to lead to environmental catastrophes? Mathematically, such consequences seem to be unavoidable. This assumes, however, that current trends will continue indefinitely, that there will be no change in behavior, and no reaction by either policymakers or people. Here it is argued that such an assumption is unrealistic, given current knowledge and current awareness of the problem by governments and people alike.[15] If appropriate steps are taken, major environmental damages can be avoided in spite of significant reductions in forest covers and tree stocks.

There are a number of reasons for this cautious optimism. The first, and perhaps most important one, is that removal of trees as such does not cause damage. Damage occurs only when trees are removed from specific locations in which they protected the soil against intense sunshine, wind, and water erosion. Whether such damages occur in a given location depends on soil characteristics: the slope of the land; the intensity, magnitude, and duration of rainfall; wind velocities; the duration of dry and wet cycles; and the types of plant cover during various seasons. Many of these variables are subject to control by man. Sloping land can be protected not only by the roots of trees, but by contour plowing (if subsequently used for agriculture), by terracing, by grassy strips or hedgerows planted at regular intervals, and by gully check dams that reduce the velocity of water runoff (Schramm 1979). Damage from the wind can be greatly reduced or eliminated by the planting of wind breaks (see chapter 10), and the excessive siltation of reservoirs can be prevented by the design and protection of sufficiently large shelterbelts in the upstream watershed.[16]

Some large, contiguous areas must be fully protected if severe damages are to be avoided. The most urgent concerns are for tropical forest areas that contain large numbers of unique plant and animal specimens (see chapter 5); key mountain watersheds, such as the Aberdares and Mt. Kenya mountains in Kenya that protect the headwaters of many major rivers; and regions where removal of tree cover would destroy the thin layer of top soil, leaving the land barren and infertile. Wide areas of the Amazon River basin fall into this group (Browder 1985). Protection must also be provided for Africa's unique wildlife parks and ranges, which are a priceless heritage not only for Africans, but for the world as a whole.

Outside of these sensitive areas, however, the question of whether 10, 20, 50 percent or none of the land should be retained under forest cover in a given region is a question that can be decided only on the basis of informed judgment about the importance of and benefits from retention as opposed to utilization or conversion to other uses. Nevertheless, if exploitation is found to be acceptable, the manner in which it proceeds can be of critical importance for future land use and the protection of the land's fertility and recuperative capacity. Hence, it is important to reduce or eliminate the current, indiscriminate destruction of natural forests. The ways in which wood is obtained and the slash-and-burn methods of agriculture need to be systematically replaced by less harmful practices, such as those discussed in the following section. Subject to these caveats, the utilization of existing forest resources to satisfy the growing demands for energy by rural and urban populations as well as industrial users is in the best interest of most African countries at their present stage of economic development. It will minimize energy costs to local industry as well as to those groups of society that can least afford it. Exploitation will utilize a domestic resource, minimize imports, and provide domestic employment. Even if exploitation depletes existing stocks, it provides a breathing space in which scarce resources of foreign exchange and capital can be used to sustain economic development, rather than to pay for the import of alternative energy supplies.

Optimizing Production and Consumption of Wood Fuel

In most countries, the major sources of wood fuel supplies for urban and industrial users, roughly in order of importance, are natural forests, land clearing operations, woody savanna and range lands, private lands, and man-made plantations. Obviously, the importance of the respective sources varies by location. Table 9-5 presents an example from Kenya, where recent studies have provided a more detailed identification of sources than is usually available elsewhere in Africa. In Kenya land clearing is by far the most important source of fuelwood at present, accounting for about 80 percent of total supplies, followed by plantation smallholder woodlots and miscellaneous sources. Sooner or later the supply of wood from land clearing will be exhausted, and alternative sources or alternative fuels will have to be substituted.

In larger cities, charcoal is by far the most important household fuel. Because of its higher energy density per volume (approximately 1.9 times that of wood), it can be transported over longer distances than wood. Also, in densely populated urban areas, smokeless charcoal is much preferable to smoky wood, which requires either outdoor cooking or the use of elaborate and expensive woodstoves equipped with appropriate chimneys. The wood fuel used for industrial purposes—

Table 9-5. Sources of Charcoal for Two Major Urban Markets in Kenya

Urban market	Source of supply	Type of resource[c]	Sustainability[a]	Average distance to markets (kilometers)	Market share[b] (percent)
Nairobi	Aberdares	Wattle plantations[c]	A	20–200	6
	Ukambani	Rangeland clearing	C,D	80	15
	Mau/Narok	Forest and range clearing	D	150	20
	Mtito Andei	Forest and range clearing	B,C,D	200	10
	Laikipia	Forest clearing	D	220	20
	Laikipia	Eucalyptus plantations	A	220	6
	Baringo	Rangeland clearing	C	240	15
	Others				10
Mombasa	Kwale	Forest and range clearing	B,C,D	50	40
	Kilifi	Range clearing	B,C,D	70	30
	Malindi	Forest and range clearing	B,C,D	100	15
	Taita	Range clearing	C	140	10
	Others				5

a. Sustainability index: A, sustainable smallholder and plantation production; B, sustainable low-intensity production; C, nonsustainable low-intensity production; D, nonsustainable salvage and clearing operations.

b. Seasonal variations in price and accessibility change these market shares.

c. Wattle, an acacia species, is grown for use of its bark in the tanning industry. Most of it is grown on smallholder woodlots of between 20 and 100 square meters. Charcoal production is a by-product.

Source: UNDP/World Bank data.

especially tobacco curing, tea processing, and brick making—is almost exclusively firewood. By their very nature, however, these industries are in rural areas and do not necessarily compete directly with wood fuel supplies to urban areas.

Optimizing strategies that could be adopted to utilize the dwindling resource base should include

- Systematic introduction of improved kiln technology to reduce losses when wood is converted to charcoal
- Minimization of transport costs to markets, thereby extending the potential areas of supply
- Minimization of fuel losses resulting from charcoal breakage
- Energy conservation through the use of more efficient appliances or furnaces in households and industry
- Land-use regulation and effective enforcement that would prevent overcutting or encroachment on environmentally sensitive areas
- Increased collection of revenue from wood fuel production to finance an effective forest management and advisory service as well as to provide roads, training, and other services to wood fuel producers
- Improvement in the reliability of supply of alternative fuels such as kerosene, LPG, or electricity, but with prices adjusted to fully reflect their full economic costs to the economy.

Medium- to long-term strategies to extend or replace existing wood stocks to meet future demands should consist of the following:

- Establishment of a unit to monitor markets and update forecasts of fuel demand by region and user group, in ways that would take into account projected prices as well as availabilities of substitute fuels
- Regular monitoring of changes in remaining accessible forest stocks through spot surveys and satellite photography
- Systematic identification of public, communal, and private lands that have opportunity costs and would be suitable for the establishment of future fuelwood plantations
- A systematic program, backed by technical advisory services and appropriate credit facilities, to support and stimulate the establishment of private fuelwood plantations and wood lots to meet projected future demands that could no longer be satisfied from remaining resources.[17]

Overall, the immediate objective would be to increase as quickly as possible the efficiency of ongoing wood fuel production and charcoal conversion, as well as the efficiency of ongoing wood fuel use, subject to appropriate benefit-cost criteria. For the longer term, the objective would be to replenish the dwindling resource base by providing access to more remote locations with unused or underutilized wood resources and by creating man-made supplies from plantations, timed for harvesting when the existing resources become so scarce that market prices rise sufficiently to recover the costs of wood replacement.

This longer-run strategy involves some risks. Projections of supply, demand, and prices will have to be made at the time of planting, when market prices presumably are well below replacement costs. After the investment in new plantations has been made, changes in the price and availability of alternative energy resources (such as kerosene or natural gas) could make them more attractive to consumers than wood fuels from woodlots or plantations planted long ago. Judicious decisions and acceptance of risk by governments and foreign donors, who currently are the major financiers of new investments, would be needed.[18] For this reason careful forecasting of long-term supplies and demand, as well as of future costs and prices, is important. The longer-run strategies should be adopted only if it can be shown that fuel resources from such plantations will have a high likelihood of being less, or at least not more, costly than those of properly shadow-priced alternative sources of energy.

The immediate strategy of improving the efficiencies of production and use is less risky. Today, almost all of the charcoal in Africa is produced by traditional earth kilns with conversion efficiencies from wood to charcoal, figured by weight, of 10 to 12 percent. Although much higher rates of conversion have sometimes been achieved, this requires a skill that only a few charcoal burners have. In contrast, the use of brick or steel kilns results in average efficiencies of between 28 to 35 percent, or about two and a half times higher than when earth kilns are used. Improved earth kilns, such as the Casamance or the Subri Fosse kilns, achieve somewhat lower conversion rates of 20 percent or more, but they are more flexible and less costly to build and move around than the brick and steel kilns. Clearly, the systematic use of such kilns would more than double the useful output of charcoal from a given volume of wood.

These improved technologies are not now being used for several reasons. Although widespread in South America, they are new to Africa. Outside agencies, such as the U.S. Agency for International Development (USAID) or Gesellschaft für Technische Zusammenarbeit (the German Technical Aid Organization, or GTZ), have long assisted African countries in adapting technology to particular circumstances. This was always done within the context of small-scale pilot projects, however, with no resources available for widespread introduction and dissemination.[19] Charcoal production is undertaken by myriad small-scale, often part-time charcoal burners, who are generally illiterate, have no financial resources of their own, and no access to credit. Without systematic outside help and systematic dissemination, these producers would not and could not adapt to these new technologies.

A second important part of this strategy is the need to develop and maintain proper utilization and conservation practices on existing and maintainable forest lands. At the same time, enough revenues need to be raised from the exploitation of forest resources to cover the costs of effective forest management and advisory services. At present, most of the forest lands in Africa are the common property of villagers or tribes or owned by the government. Well-defined user rights are lacking. As a result, wood cutters or charcoal burners have no interest in maintaining or conserving the existing resource base. Although stumpage fees may be levied by either local or central authorities, they are widely avoided because of the lack of proper policing. What is needed instead is the establishment of contractual user rights for individuals or communities, with stumpage fees or lease payments based on a realistic assessment of the netback (that is, residual resource) value of the standing wood, and with enforceable rules for cutting regimes.[20] If such leases or cutting rights are properly designed to provide the wood producer with a reasonable source of income, that producer will, in turn, try to protect the resource base against encroachment by others, thereby substantially reducing or eliminating the need for public protection of the forest.[21]

A Case Study: Malawi

Malawi is a small country in southern Africa with a population of around 7 million people. Forests play a vital role in the economy, providing approximately 90 percent of the nation's domestic and industrial energy requirements and a substantial volume of timber (World Bank data). Malawi's forests also provide recreational facilities and significant environmental benefits. The major consumers of wood are presented in table 9-6. The rural population uses fuelwood for cooking and heating as well as poles for building, and accounts for around 60 percent of the total demand. The principal source of energy for urban households is charcoal, which is made from indigenous wood using very inefficient conversion methods. The tobacco and tea industries, Malawi's chief sources of foreign exchange earnings, are critically dependent on large supplies of fuelwood for curing and barn construction. In addition, informal village industries use wood for a variety of purposes, including brick making, smoking of fish and meat, beer brewing, and lime burning.

As a whole, Malawi's total forest resource is still extensive. Approximately 38 percent of Malawi's land area is currently under forest cover (Interdisziplinäre Projekt Consult 1988a). This aggregate picture, how-

Table 9-6. *Estimated Wood Consumption in Malawi*

User	Millions of cubic meters	Percent
Rural households	5.1	60
Urban households	1.0	11
Tobacco and tea estates	2.0	23
Village industries	0.4	5
Urban services and industry	0.1	1
Total	8.6	100

Source: World Bank data.

ever, hides very significant regional imbalances in supply and demand. The basic problem is that the population is concentrated in the south and central regions, while the bulk of the forests are in the relatively underpopulated north (see table 9-7). Because of the high costs of transportation, northern surpluses of conventional firewood cannot be used to fill deficits in the south.

In the south and in much of the central region there is now a substantial gap between the demand for fuelwood and its sustainable supply from customary woodlands. This deficit is being met by depletion of the stocks, which further widens the gap between demand and the sustainable flow of wood and leads to ever more rapid destruction of stocks. At the same time, demand for wood energy is increasing rapidly because of population growth of around 3.7 percent per year, compounded in recent years by an influx of several hundred thousand refugees from Mozambique. Urban population growth of over 6 percent a year further increases demand because wood consumption per capita is higher in urban areas (1.4 cubic meters per year) than in rural areas (0.8 cubic meters per year). Deforestation is exacerbated by widespread land clearing for agricultural purposes as a result of population pressure and

Table 9-7. *Distribution of Population and Forest Resources by Region, Malawi, 1987*

Item	North	Central	South
Population (millions)	0.91	3.11	3.96
Forests (million hectares)			
Customary land	1.64	0.91	0.50
Forest reserves	0.20	0.33	0.29
Government plantations	0.05	0.02	0.02
Private plantations	n.a.	0.01	0.01

n.a. Not available.
Source: Interdisziplinäre Projekt Consult (1988a).

the traditional practice of shifting cultivation. Under current trends it is estimated that the indigenous forest on customary land will have totally disappeared in the south by 1993 and in the central region by 2003 (Interdisziplinäre Projekt Consult 1988a).

Deforestation causes soil erosion and damage to watershed areas, including silting up of water courses and flooding, that would have long-term negative and possibly irreversible implications for agricultural productivity and for economic growth and welfare. Destruction of the forest resource would also have major consequences for the future energy supply of the country. At present it is unlikely that alternative energy sources will become available at prices the population in general could afford. There are no known deposits of oil or natural gas in Malawi. Imported fossil fuels are expensive because transportation costs to landlocked Malawi have increased several-fold in recent years with the disruption of direct routes through Mozambique. Rural electrification is unlikely to materialize in the near future because of the high establishment costs, and, solar energy and biogas are still at the experimental stage. Limited coal deposits of poor quality occur in the far north of the country but mining and transportation costs are high. Therefore development of a wood energy program that can meet the future need for fuelwood and building poles, while at the same time conserving the environment, is of critical importance for Malawi.

Evolution of a Strategy for Wood Energy

Over the past decade Malawi has begun to address the emerging fuelwood deficit and the increasing destruction of indigenous forests. A strategy has evolved from an initial focus on direct tree planting by the government to a comprehensive and multifaceted program for the management of wood on a sustainable basis. In particular, the role of the government has gradually shifted toward the creation of an institutional and policy environment that encourages private production and conservation and more efficient use of existing resources.

The initial response to deforestation was for the Forestry Department to establish fuelwood plantations close to Blantyre, Zomba, and Lilongwe, the main centers of population. Public tree planting seemed appropriate because experience elsewhere showed that by the time wood becomes so scarce that people start to plant trees, environmental damage has already occurred. Individuals are unlikely to take timely actions to conserve wood and replant forests because even where there is a substantial gap between demand and sustainable supply, the deficit is met by depletion of stocks and the shortage of wood is not perceived. By the time the scarcity begins to be seriously manifested, much of the resource base may have disappeared. Even when the scarcity has been recognized and reforestation efforts initiated, depletion of the natural forest will continue because of the considerable lag between tree planting and harvesting.

In addition, individuals fail to conserve or produce wood because there is no private ownership of trees or at least no well-defined and enforceable system of user rights. In Malawi, as in much of Africa, forests are traditionally considered to be openly accessible, common property. Wood for domestic purposes is free, while wood for commercial purposes is subject to a very low stumpage fee that is rarely collected. In the past, when the sustainable yield from forests greatly exceeded demand, wood could in some ways be considered a quasi-public good: as consumption by one household did not reduce the benefits derived by the rest of the community, and the marginal social cost of additional consumption was zero. Today, demand for wood exceeds sustainable supply in many areas. The policy of free wood now implies a considerable divergence between the private and social costs of wood consumption because the price of fuelwood to consumers reflects only the low private costs of cutting the wood rather than the significant long-term social cost of replacement. Raising the price of wood closer to replacement costs would do little to increase the incentives for private tree planting and wood conservation if collection of the stumpage fee is not enforced. Forests have another characteristic of public goods—it is difficult to exclude consumers who do not pay. Public provision of these types of goods is often considered appropriate.

Direct production of wood by the government, financed through the central budget, was originally adopted to deal with the fuelwood problem in Malawi. Between 1980 and 1987 approximately 15,000 hectares of peri-urban fuelwood plantations of fast-growing, exotic species such as eucalyptus were established by the Forestry Department. The yields from these plantations are very disappointing—the average mean annual increment (MAI) is about 4.7 cubic meters per hectare per year against a potential 10 to 14 cubic meters per hectare per year (Malawi Forestry Department and Forestry Research Institute of Malawi 1987). In some extreme cases, the MAI of the plantations is lower than that of the unimproved miombo woodland that was cleared from the site (Hardcastle 1988). This is partly because plantations were established in unsuitable locations and with inappropriate species and provenances. In addition, insufficient attention was paid to tending and weeding in the first years after planting and inadequate fire protection was provided. Although wood yields are low, the per hectare costs of establishing the government plantations were high owing to heavy outlays on vehicles, staff housing, offices, roads, and other infrastructure. The ultimate financial cost per cubic meter of wood produced is therefore extremely high—approximately 21.9 kwacha per stacked cubic meter (in 1986–87 prices), which is about eight times higher than the stumpage rate of 2.7 kwacha per stacked cubic meter that prevailed in 1987 (World Bank data). The economic cost of wood produced on peri-urban plantations would be even higher if the opportunity cost of the land utilized was included.

The other main government initiative in the early 1980s was the establishment of a network of retail nurseries to sell seedlings at subsidized prices to smallholders and estates for tree planting. Sales from the nurseries were low. Inadequate extension services to smallholders for tree planting may partially explain this, but planting trees was found to be well accepted and practiced where a nearby cash market for poles existed (Malawi Ministry of Forestry and Natural Resources 1982). This suggests that farmers did not find it financially attractive to grow trees for fuelwood given the low producer prices for wood and the availability of free wood in the virtually uncontrolled and unprotected natural forest.

These early initiatives did little to slow the destruction of the natural forest in deficit areas. The efforts were, however, part of a learning process that has contributed to the evolution of a broad-based strategy for the management of wood energy. The experience with plantations demonstrated that it is technically and fiscally unfeasible and economically undesirable for the Forestry Department to produce fuelwood. There are severe shortages of qualified manpower in the Forestry Department. Planting even 15,000 hectares over the last few years overwhelmed the capacity of the staff to implement the program effectively, yet the annual output from these plantations represents less than 1 percent of total annual demand. In addition, Malawi is currently undergoing severe cuts in public expenditure, and budgetary constraints prevent the Forestry Department from planting on a sufficient scale to make a substantial contribution to sustainable supply. Such planting would in any case be inefficient because the government has proved to be a high-cost producer of wood. Wood can be produced at much lower cost by the private sector, particularly by smallholders (see table 9-8). The poor sales of seedlings from the retail nurseries demonstrated, however, that farmers will not produce wood without adequate financial incentives.

To address these problems the government is now implementing a comprehensive package to promote tree planting by both smallholders and commercial users. It includes the provision of subsidized seedlings and extension services, incentive payments for tree planting, pricing reform (including substantial increases in wood prices), and increased protection and revenue collection in the natural forest. In addition, the government is exploring ways to use existing low-cost sources of fuelwood such as thinnings and trees past their prime from industrial

Table 9-8. *Present Discounted Value of Total Economic Costs of Establishing One Hectare of Trees in Malawi*

Item	Kwacha
Government plantations	894
Small-farmer woodlots	258
Fiscal cost to government of small-farmer woodlot[a]	198

a. Subsidies to seedlings, transportation, extension, monitoring, and tree planting incentive payment.
Source: World Bank data.

plantations and waste wood from land clearing operations. The program also includes several components to conserve wood through the introduction of simple but more fuel-efficient technologies for tobacco curing, charcoal making, and domestic cooking stoves.

Increase in Stumpage Rates

The very low (and rarely enforced) stumpage rate constitutes a serious market distortion and is a major cause of the rapid depletion of the natural forest in Malawi. The government is now implementing a phased increase in stumpage fees for wood cut for commercial purposes (tobacco curing, village industries, charcoal making, resale to urban households, and so on) with a view toward reaching replacement cost over a period of ten years. This increase in price is essential for managing both the supply of and the demand for wood energy because it sends a signal to the private sector that wood is not a free resource. It should promote the adoption of more energy-efficient technologies by urban households and industry and create incentives for private production of wood. In addition, the price increase will generate revenue for the government that can help finance other essential elements of the fuelwood strategy, such as improved forest protection, subsidized seedlings, tree planting incentives, extension, and research.

Wood cut for domestic purposes will continue to be free of charge for several reasons. First, there has always been a right to free wood for domestic use in Malawi, and politically it would be extremely difficult to change this. Second, the bulk of rural households are subsistence farmers who are largely outside the cash economy and would be unable to pay for the wood they need to survive. Third, it would be excessively costly, if not impossible, to enforce payment of stumpage fees for small volumes of wood extracted daily by millions of rural households. Since rural households account for the bulk of wood utilized, however, other measures are necessary to promote self-sufficiency in fuelwood for this group.

Improved Revenue Collection

Although the increase in stumpage fees is a critical part of the wood energy strategy it will have no impact if payment of fees is not enforced. Therefore the government is also improving the system for protecting the forest and collecting revenue for wood taken from forest reserves and customary lands (Malawi Ministry of Forestry and Natural Resources 1986b). The number of foresters, forest guards, and patrolmen is being substantially increased. Part of their duties is to collect stumpage fees, primarily from the major commercial users. Fee collection is backed up by spot checks and a few strategic checkpoints on roads leading to the main markets. This system should lead to significant increases in revenue; at present only about 10 percent of commercial wood is actually paid for.

Protection and Management of Indigenous Forest

In addition, the forestry staff protects forests in areas that have been heavily depleted or are environmentally fragile. Commercial cutting has been restricted or prohibited in those regions, and major users are directed to environmentally safe sources of wood, such as government fuelwood and industrial plantations, forest reserves with surplus yields, or waste wood from land clearing activities. Forestry staff educate local inhabitants on the sustainable utilization of the indigenous forests. They advise on improved silvicultural techniques including cutting practices, enrichment plantings, and fire control, in order to ensure regeneration and optimal yields over time.

Direct Promotion of Tree Planting by Farmers

The government is indirectly encouraging commercial users to plant and conserve trees by gradually increasing stumpage fees, improving revenue collection, and restricting or prohibiting cutting for commercial purposes in depleted or environmentally vulnerable areas. Rural households, which account for more than half of the demand for fuelwood, do not have to pay stumpage fees. Because of this the government is also implementing a series of measures to promote tree planting directly, with the objective of making smallholders self-sufficient in wood energy. These measures include provision of subsidized seedlings from retail nurseries, extension services, incentive payments for tree planting, and systematic public education about the importance of tree planting. Seedlings are sold at approximately one-fifth of their production cost. In the past, anticipated demand for seedlings by farmers did not fully materialize. With the additional incentive payment for tree planting and the gradual increases in the stumpage rate for commercial wood, it is expected that farmers will respond more favorably. In addition, extension units are being set up on a pilot basis in areas of acute fuelwood shortages. Extension staff will give farmers simple technical advice on tree planting, harvesting, and wood usage and register them for the tree planting incentive payments. They will also assist the forestry staff working in the indigenous forests in teaching the local people how to utilize the natural resource on a sustainable basis.

The tree planting incentive scheme evolved from the initial experience with the nurseries, which made it clear that farmers had little financial incentive to plant trees even when provided with seedlings at subsidized prices. Farmers face a choice between using their land and labor for planting trees or agricultural crops. Growing trees requires tying up land for many years, with the costs incurred principally in the first year and the benefits occurring only later. Small farmers also tend to have high private discount rates. These factors, combined with the current low stumpage rates and widespread free access to wood, mean that financial returns to fuelwood production are very low or negative. To compensate for the impact of price distortions on the profitability of establishing woodlots by individual farmers, an investment incentive is required. Under the tree planting incentive scheme, farmers will receive payments for each tree that survives for two years, after which trees require almost no maintenance. The payments should make tree planting financially attractive relative to agricultural crop production on marginal land. Over time, as the stumpage rate for wood increases and restrictions on cutting in the indigenous forest are implemented, the need for the tree planting subsidies should decline. In the short run, however, production subsidies are economically justifiable as a way to correct the existing market distortion that discourages private production.

Encouraging smallholders to plant trees through provision of subsidized seedlings and incentive payments is a much more cost-effective way for the government to promote wood production than its earlier approach of establishing plantations (see table 9-8). Farmers can produce fuelwood at approximately one-third of the total economic cost of wood from government plantations, and the fiscal burden on the government is much less. The financial cost to the government of the subsidies for each hectare planted by farmers is only about a quarter of the cost per hectare of Forestry Department plantations.

The actions on wood pricing and revenue collection indirectly create incentives for greater wood conservation. In addition, initiatives to develop and disseminate more fuel-efficient technologies in specific areas of high wood consumption are expected to reduce the demand for wood.

Tobacco Curing Energy Conservation Program

Tobacco curing. The flue-cured tobacco industry is extremely important to the Malawian economy because it generates a large share of export earnings. The industry also utilizes the bulk of commercial fuelwood extracted from the natural forests and has caused severe deforestation in many areas.[22] In the past, the use of wood in tobacco curing has been very inefficient. A 1984 survey of estates found that the national average for fuelwood consumption in the tobacco curing industry was 42 stacked cubic meters per ton of tobacco cured (Malawi Ministry of Forestry and Natural Resources 1986a). There is substantial potential to reduce the amount of wood burned per barn load of tobacco cured. Research trials have demonstrated that wood consumption can be reduced to between 10 and 15 stacked cubic meters per ton by using an improved furnace and better management and curing techniques. This reduction does not require large investments—in fact the new slot furnace is significantly cheaper, as well as more efficient, than the conventional grid furnace. Installation of the improved equipment will not by itself bring about the desired result; improved management and curing practices are found to be the key to reducing the amounts of fuelwood needed.

The main focus of the program is now on disseminating the research findings and training estate managers, furnace builders, and furnace stokers/operators in the new techniques. Monthly newsletters are sent to the tobacco estates and demonstration field days, workshops, and seminars are being held, together with visits to individual estates for on-the-job training. Although many of the large estates have already begun to adopt the new techniques, it will take time to reach the 2,000 or so small estates that are the most inefficient. If eventually the entire crop of Virginia tobacco were to be cured using the improved techniques, average wood consumption would drop from 42 to 12 stacked cubic meters per ton. This would result in a saving of around 600,000 stacked cubic meters per year, or the equivalent of clearfelling 10,500 hectares of natural forest each year. Trials are also under way to assess the use of pine charcoal as a substitute curing fuel, which would further reduce destruction of the natural forest.

Improved household stoves. Under the pilot stove component of the program, a cooking stove has been developed that uses pine charcoal, which can be produced without destroying natural forest, and that can

Table 9-9. *Cooking Trials of Traditional and Improved Stoves*

	Average consumption (kilograms)	
Type of charcoal	*Traditional stove*	*Improved stove*
Pine	2.10	1.18
Indigenous	3.15	2.23

Note: Five hundred grams of beans were cooked in the trials.
Source: Interdisziplinäre Projekt Consult (1988b).

achieve fuel savings of between 30 and 45 percent (see table 9-9). Local artisans can easily produce the stoves using old oil drums for the metal casings and local clay for the ceramic liners. Production costs are low (around US$5), and at current charcoal prices the pay-back to stove purchasers is less than three months (Malawi Ministry of Forestry and Natural Resources 1987). Two pilot production centers are operating successfully, and stoves are now being sold in local hardware stores. A sample of households have tested the improved stoves in their homes, and preliminary indications are that consumers are reacting very positively to them. The next stage is to move from pilot production, which has been run with technical assistance from Kenya, to full-scale commercial operations without government subsidies or assistance. It is anticipated that within three years some 45,000 improved stoves will have been produced and sold in the Blantyre and Lilongwe urban markets. The savings of wood will be equivalent to clearfelling about 1,000 hectares of indigenous forest annually and will be even more if households switch to using pine charcoal.

Pilot program for charcoal production.   Another energy conservation component of the program is the introduction of more efficient methods for making charcoal. At pilot charcoal production sites beehive brick kilns convert wood into charcoal at an efficiency rate (by weight) of about 32 percent under normal operating conditions, which is about three times higher than the conversion rate of the traditional earth kilns. The brick kilns are built from local materials and are operated at very low cost by local labor assisted by oxen, with minimal requirements for training, capital, or foreign exchange. The production centers are located at the 55,000-hectare pulpwood plantation in the Viphya mountains in the north and at an old timber plantation in the south.

There are extensive plantations in Malawi that were intended to supply wood for domestic wood processing and for export-oriented pulp and paper industries. The anticipated market for pulp and timber products never materialized, however, and many of the trees in those plantations are now past their peak and are no longer suitable for industrial purposes. Pulp and timber schemes require regular silvicultural treatment, which generates a steady flow of wood from forest thinning that until now has been wasted. Because most of the thinning activity occurs in the Viphya, far from the major demand centers for fuel (see figure 9-1), the most economic way to use the waste wood is to produce charcoal, which minimizes transportation costs.

The charcoal component illustrates how the management of wood energy in Malawi continues to evolve. Various crisis scenarios for Malawi tended to focus exclusively on the gap between the demand for fuelwood and the sustainable supply from the indigenous forest in the southern and central regions; other possible sources of fuelwood were more or less ignored. The initial strategy therefore emphasized wood production and conservation. Under the more recent charcoal program alternative significant and low-cost sources of fuelwood have been identified. Although additional tree planting is clearly part of the solution to the fuelwood problem, a strategy for a least-cost energy supply also calls for better use of the existing potential: thinnings from industrial plantations, including clearfelling of trees too old for industrial purposes; wood from land clearing for agricultural expansion; and ecologically sound and sustainable exploitation of forest reserves in wood surplus areas in the north (see figure 9-1).[23]

Preliminary results from the charcoal component indicate that charcoal made from surplus wood from industrial plantations could meet a substantial part of the commercial demand for wood energy. This charcoal can be produced at a lower cost than the wood from peri-urban government fuelwood plantations (see table 9-10). It is estimated that the annual flow of wood from thinning plus that provided by clearfelling of old trees could supply more than 100,000 tons of softwood charcoal a year for at least the next ten years (Interdisziplinäre Projekt Consult 1988a).

On the demand side, trials undertaken on tobacco estates during the 1987 tobacco curing season demonstrated the technical feasibility of using charcoal rather than wood in curing. Large-scale trials in 1988 were designed to measure fuel use and finalize the necessary modifica-

Table 9-10. *Landed Costs of Charcoal from Different Sources*
(kwacha per landed tonne of charcoal)

Outlet	Viphya thinnings	Dedza wood waste[a]	Lilongwe peri-urban eucalyptus	Blantyre peri-urban eucalyptus
Lilongwe	125	90	147	—
Blantyre	165	113	—	160

— Not applicable.
a. Dedza is a 6,800-hectare timber plantation in Central region.
Source: Interdisziplinäre Projekt Consult (1988b).

Figure 9-1.   *Malawi: Main Transport Routes for Viphya Charcoal*

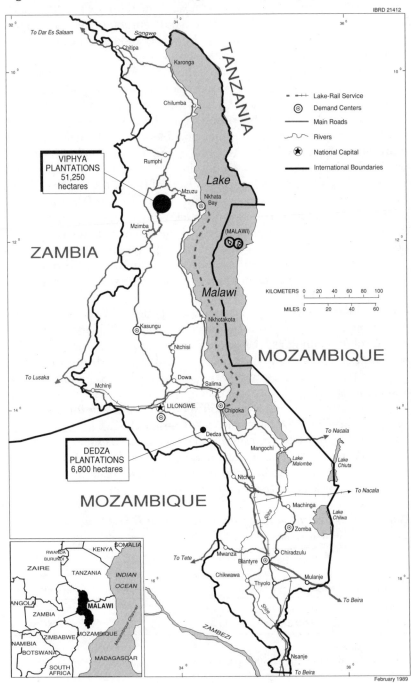

tions in the conventional design of tobacco barns. The market potential for softwood charcoal in tobacco curing is currently estimated at about 20,000 tons, on the assumption that about 40 percent of the estates have serious fuel supply problems and would welcome an alternative fuel. If around 40 percent of the total Virginia tobacco crop were to be cured using softwood charcoal this would reduce the annual clearfelling of natural woodlands by 2,000 to 6,000 hectares.

Current hardwood charcoal consumption by urban households is around 50,000 tons per year (Malawi Ministry of Forestry and Natural Resources 1984). Although consumers prefer hardwood charcoal to light pine charcoal, the government recently prohibited charcoal making in the indigenous forests. With diminishing supplies of hardwood charcoal to the cities and the introduction of stoves that use softwood charcoal, pine charcoal is gaining acceptance. The entire 50,000 tons per year can therefore be regarded as the potential urban market for softwood charcoal. At present approximately 9,000 hectares of woodlands are clearfelled every year to supply charcoal to households in the cities. If softwood plantation charcoal replaced all indigenous hardwood charcoal for household use, defores,tation would be reduced substantially.

The next step is to move from the pilot production stage, under which pine charcoal is currently produced and distributed by the Forestry Department with significant outside technical assistance, to full-scale commercial production and marketing by private sector operators without government intervention or subsidies. The government intends to offer concessions to private producers of charcoal, who will use waste wood from timber plantations and be charged appropriate stumpage fees. Although there remain a number of technical issues to be resolved, particularly on the transportation side, there appears to be considerable interest in the private sector in taking up concessions, including tobacco estates. Small-scale entrepreneurs are interested in the household market and large-scale entrepreneurs in production and distribution to the industrial market.

Conclusions

The current strategy for the management of wood energy in Malawi involves the simultaneous implementation of a comprehensive package of policies and programs. Few, if any, of the individual elements can have an impact alone—each is mutually dependent on and reinforces the impact of the other components. Thus the fuel-efficient technologies being developed might not be adopted without the financial incentives for energy conservation created by increases in stumpage rates and improved revenue collection. Conversely, in the absence of low-cost techniques to conserve wood increased stumpage fees would merely be inflationary and cause hardship to consumers. Restrictions on commercial cutting and prohibition of charcoal making in heavily depleted and environmentally fragile areas are essential to prevent the total destruction of Malawi's indigenous forest. These actions must, however, be complemented by efforts to provide environmentally safe and inexpensive energy alternatives to users. Charcoal production is a key element in this strategy because it permits immediate use of distant sources of surplus wood at relatively low cost. Although pine charcoal offers an environmentally safe, alternative source of energy, its production has to be accompanied by the development of inexpensive technologies that will enable users to switch from indigenous charcoal or wood to softwood charcoal. Reliable, low-cost distribution channels must also be developed. Cooking stoves have been specifically designed for use with pine charcoal, and the marketing and distribution of the stoves and the charcoal should be closely coordinated. Likewise, for the successful utilization of pine charcoal in tobacco curing it will be essential to develop a guaranteed regular supply of charcoal to the tobacco estates. Incentives for tree planting, subsidization of seedlings, and extension services are other essential ingredients in the long-term strategy to shift users away from the natural forest to cultivated wood sources. These efforts to promote tree planting by the private sector are in turn reinforced by the actions on stumpage fees and revenue collection.

Although the difficulties inherent in coordinating such a complex program should not be underestimated, the strategy represents one of the most comprehensive and promising attempts to develop a sustainable, low-cost, and environmentally sound supply of energy in a developing country with low income. The program also provides a good example of an effort to minimize the burden on the government budget. Directors of development programs in all sectors in Malawi must take into account the constraints on public expenditure, particularly for financing recurrent costs of projects for which donor funding has ended. The earlier focus on Forestry Department fuelwood plantations is placing a severe strain on the government budget because it involves significant operating costs, with limited scope for recovering the high costs of the wood produced. A critical requirement for a successful wood energy program is that it should be fiscally sustainable. Under the new program, substantial government income should be generated by the increase in stumpage fees and improved revenue collection. In addition, granting concessions for private charcoal production to use wood thinned from industrial plantations will generate revenue from stumpage fees as well as permit the government to save money that is currently being spent on essential maintenance of the plantations. These increased revenues will help finance other elements of the forestry program such as research, extension services, provision of subsidized seedlings, and incentives for tree planting.

Naturally, some aspects of the program are specific to Malawi. Much of the approach does, however, appear to be of relevance for other countries trying to develop a wood energy strategy, in particular, the importance of a comprehensive strategy to promote wood production, conservation, and better utilization of existing resources by the private sector. This strategy uses a variety of direct and indirect instruments, focuses on development of a least-cost supply strategy, and requires any program to be fiscally sustainable.

Notes

1. Although this paper addresses the issues related to the use of wood as an energy resource, it must be remembered that in most of Africa today far more wood is cut down in land clearing operations than is used as fuel.

2. In Kenya, for example, it was estimated that in 1985 urban charcoal demands provided some 30,000 full-time jobs for charcoal burners, another 400 for transporters, and 800 for retailers, generating rural incomes of about US$22 million and retail sales of about US$52 million (UNDP/World Bank data).

3. In Kenya, it has been estimated that land clearing accounts for around 80 percent of the current urban supplies of charcoal (UNDP/World Bank data).

4. Commercial energy sources include petroleum fuels, natural gas, coal, and electricity. Labeling these energy resources as commercial, in contrast to the so-called traditional wood fuels, is misleading because wood fuels for urban and industrial users are traded almost exclusively on a commercial basis.

5. Even oil-rich Nigeria is a significant importer of certain types of petroleum products, because it cannot afford the high cost of developing a local refining industry capable of supplying all of its domestic demands (UNDP/World Bank data).

6. In Malawi, tobacco curing accounts for close to 30 percent of the total amount of fuelwood consumed, which, in turn, accounts for about 80 percent of the total amount of energy consumed in tobacco curing (World Bank data).

7. Promotion of electric cooking would generally be undesirable for an electric utility because of the low load factor of 10 to 12 percent involved (at an average of two to three hourse of cooking per day). The high costs of additional generating capacity needed to serve this type of load would make cooking either prohibitively expensive (if users pay capacity charges) or would require substantial cross-subsidies from other users. Electric cooking, therefore, should be promoted only in countries with a temporary surplus of generating capacity (Ethiopia, Tanzania, Zambia, and Zaire, for example), and in those countries should be accessible only to users already connected to a grid.

8. The incremental costs of natural gas for domestic use in cities without the need for space heating are estimated by the World Bank to range from US$8 to US$11 per million British thermal units.

9. Because LPG is a by-product of the refining process for crude oil, its availability is limited to countries with domestic refineries. Only a small amount of LPG is produced compared with the quantity of crude refined, and the high transport costs of pressurized containers for LPG make its importation costly.

10. Nigeria consumed some 1.5 million tons of kerosene in 1981, of which almost 0.4 million tons were imported (UNDP/World Bank data).

11. Given the rather shaky assumptions of the underlying data sets these percentages should be considered to be estimated orders of magnitude rather than accurate projections.

12. With the present lower prices for petroleum on the world market, these percentages have declined, but are still formidable, particularly for locations in the interior of countries, where costs of transportation from ports of entry can more than double the costs at points of delivery (Schramm 1986).

13. The potential impact is illustrated by the case of Malawi. Replacing the approximately one million cubic meters of solid wood equivalent consumed by urban households in 1983 would have required additional kerosene imports of approximately 55,000 tons, which would cost about US$14 million in border prices (in January 1986). This would have increased total petroleum imports by about 40 percent, absorbed an aditional 5 percent of the country's export earnings, and increased the balance of payments deficit by 18 percent (Malawi Ministry of Forestry and Natural Resources 1984; Malawi Government 1986).

14. Other reasons for not converting to kerosene are the lack of proper appliances and personal habits and taste. Many housewives refuse to cook with kerosene because food then lacks the flavor imparted by charcoal or wood.

15. There have been many studies of the issues involved. An important contribution was made by the systematic UNDP/World Bank Energy Assessment Program, which during recent years has conducted detailed evaluations of the energy situation of every country in Sub-Saharan Africa, except the Republic of South Africa.

16. This, of course, raises the troublesome issue of policing and preventing encroachment. There are no easy solutions, particularly in areas of strong population pressures and land scarcities.

17. Such programs are subject to the condition that at the time of planting the current value of the total supply costs of plantations per unit of useful energy, supplied to the users' premises, has to be lower than the costs of alternative energy sources in both financial (including appropriate profit margins) and economic terms (properly shadow priced).

18. Investing in fuelwood plantations on the basis of projected costs and future prices is little different from investing in long-term hydropower plants with construction horizons of six to eight years or more.

19. In recent years, the World Bank has started to move strongly into this area, with projects in a number of countries, including Ethiopia, Madagascar, Malawi, and Tanzania.

20. Rules include the minimum diameters of trees that have to be left uncut to assure regeneration and optimum yield over time, enrichment plantings, the promotion of natural regeneration, and protection through firebreaks. Re-

cent studies in Tanzania and elsewhere have shown that such low-cost measures can increase the average increment of wood production by 100 percent.

21. The introduction and enforcement of such semi-privatization schemes will be much easier in areas that are being opened up for exploitation. Where overexploitation already is endemic, lease rights may not be enforceable except at substantial costs.

22. The tea estates grow their own fuelwood and are more or less self-sufficient.

23. This approach will also require a revision of the current pricing policy under which stumpage rates are the same throughout the country. Pan-territorial pricing of wood further depletes wood stocks near the main markets and discourages use of surplus wood in the underpopulated areas in the north. The Malawi government is now considering a more flexible pricing policy that takes into account transportation costs and relative demand and supply in different areas.

References

Browder, J. 1985. "Subsidies, Deforestation, and the Forest Sector in the Brasilian Amazon." World Resources Institute, Washington, D.C. Processed.

Hardcastle, P. D. 1988. *Final Report on Research Component of Second Wood Energy Project, Malawi.* Oxford, U.K.: Oxford Forestry Institute.

Interdisziplinäre Projekt Consult (IPC). 1988a. *National Energy Master Plan Biomass Sector Position Papers.* Supply and Demand Analysis and Summary Report. 3 vols. Lilongwe.

_____. 1988b. *Solid Fuels in Malawi: Options and Constraints for Charcoal and Coal.* 3 vols. Lilongwe.

Malawi Forestry Department and Forestry Research Institute of Malawi. 1987. *A Summary of Yield Forecast for Wood Energy Phase I Plantations.* Lilongwe.

Malawi Government. 1986. *Economic Report.* Lilongwe.

Malawi Ministry of Forestry and Natural Resources. 1982. *Malawi Smallholder Tree-Planting Survey.* Lilongwe: Energy Studies Unit.

_____. 1984. *Malawi Urban Energy Survey.* Lilongwe: Energy Studies Unit.

_____. 1986a. *Malawi Flue-Cured Tobacco Use Survey.* Lilongwe: Energy Studies Unit.

_____. 1986b. *Study of Administrative Procedures and Regulations of Forest Revenue Collection and Tree Planting Incentive Payments.* Lilongwe: Forestry Department.

_____. Malawi Ministry of Forestry and Natural Resources. 1987. *Blantyre Charcoal Stoves Pilot Project: Progress Report.* Lilongwe: Forestry Department.

Schramm, Gunter. 1979. "A Benefit-Cost Model for the Evaluation of On-Site Benefits of Soil Conservation Projects in Mexico." *Annals of Regional Science* 13(2):1–28.

_____. 1986. "Regional Cooperation and Economic Development." *Annals of Regional Science* 20(2):table 1.

World Bank. 1981. *Accelerated Development in Sub-Sahara Africa.* Washington, D.C.

10

Economic Aspects of Afforestation and Soil Conservation Projects

Dennis Anderson

This paper presents a simple approach to estimating the economic bene-fits of afforestation and soil conservation projects in farming areas. The projects considered involve the planting of trees in cultivated fields, in copses, on farm boundaries, and as windbreaks or shelterbelts. Trees are planted for various purposes, including the production of wood, fruit, and other tree products, protection of crops from storm damage, and protection of soils from erosion. These practices are variously termed farm, and agro, social, or rural forestry, and can often be quite complex (as with alley cropping).

The data used in this chapter are based on work in Nigeria (Anderson 1987). But I am sufficiently encouraged by the comments received from people working in other regions to believe that the approach might have wider applicability—principally in regions where the carrying capacity of the soil is being depleted by the loss of tree stocks. The ecological pa-rameters would, of course, be different, but not the underlying method-ology. This chapter summarizes the approach with this in mind and includes some additional points about the kinds of social and scientific research needed to take the analysis forward—and, more important, to define economic policies better.

Methodology and Parameters

The approach was developed for the analysis of an afforestation pro-gram in the arid zone of northern Nigeria. Roughly, this zone comprises the upper half of the five northern states of Bauchi, Borno, Kaduna, Kano, and Sokoto and has an area of about 170,000 square kilometers. It is bounded on the north and west by the Niger Republic and on the

The author is indebted to Kenneth Sigrist for many valuable suggestions while developing the methodology used in this chapter, and to Robert Monahan for his encouraging and constructive comments in reviewing it.

northeast and east by Chad and Cameroon. The annual average rainfall varies from 200 millimeters in the northeast to 800 millimeters on the southern edge of the zone (12 degrees north of the equator), near the city of Kano. Areas around Lake Chad, it is reported, may have negligible or no rainfall in some years. The area has a harsh climate with a long dry season (eight to nine months at the southern edge, and nine to ten months at its northern edge). The estimated population is about 20 million, and there is a dense livestock population.

Pressures on tree stocks from fuelwood consumption, land clearance, and farm cattle have been appreciable for decades, and the rate of decline of tree stocks has for a long time exceeded the mean annual increment (MAI) of the remaining stocks on farmlands and woodlands in the zone (Anderson 1987). Conservative calculations that concentrate on the use of fuelwood alone (not the whole or in places even the major cause of the losses) are sufficient to make the point. Local surveys show that over 90 percent of the population uses fuelwood for cooking; at a rate of 0.7 cubic meters per capita per year, this gives a consumption rate of around 13 million cubic meters per year, which is perhaps two to three times the MAI from local forests, woodlands, and farmlands, and more than 100 times the MAI of forestry plantations (Anderson 1987). In addition, it is thought that the MAI per hectare may be declining from the loss of seedlings and saplings to livestock. The consequences, which are familiar in other parts of the world, include (1) a marked decline in farm tree stocks; (2) increased in encroachment by farmers on public forest and game reserves; (3) the harvesting of tree stocks without replenishment farther south where woodlands and forests are still abundant (firewood is transported 200 to 400 kilometers to the north, and there is much visible evidence of deforestation in the south); and (4) a threatened decline in soil fertility. The decline in soil fertility has probably occurred already in many areas, though there is a deplorable lack of good scientific data to document the loss. Gulley erosion, the loss of topsoil to wind erosion, greater surface evaporation and reduced moisture in the soil as wind velocities increase, reduced recycling of soil nutrients, and the possibility that dung and crop residues will be used as fuel if fuelwood becomes scarce all contribute to the decline of soil fertility. In addition, storm damage to crops is reported to be more frequent, especially just after seeds have been sown and have germinated; the topsoil frequently shifts, the farmers have to re-seed, and there are delays in germination.

What would be the benefits of an afforestation program? In northern Nigeria, there would be four:

- The benefits of stemming future declines in soil fertility
- The benefits of improving current levels of soil fertility

- The benefits of acquiring the tree products themselves (firewood, poles, and fruit)
- The benefits of increasing the availability of fodder. (Fodder can be increased if soil fertility is increased and if fodder trees and shrubs are planted as part of a farm forestry program. In turn, this would enhance the economic output of livestock activities.)

As far as I know, it is usual to concentrate on the benefits of tree products in the appraisal of afforestation projects. In the arid zone of Nigeria such benefits, though not trivial, would be fairly small and promise modest rates of return to investment of around 5 percent. Allowing for the improvement of soil fertility and the increased availability of fodder, however, raises the net present value of benefits fourfold and the economic rate of return to over 15 percent. Moreover, these calculations make highly conservative assumptions about the ecological benefits.

To begin the calculations, it is necessary to note two types of afforestation investment in the region, each having similar purposes and qualitatively similar ecological effects but different costs, risks, and quantitative effects. One is a program of shelterbelts planted by the forestry services. A shelterbelt is generally a linear planting of about six to eight rows of trees (eucalyptus and neem have been popular in Nigeria) that protects roughly up to 200 square meters of farmland on the leeward side. A typical arrangement is a series of one or two dozen shelterbelts from five to ten kilometers long, though arrangements vary. Thus, ten-by-twenty-kilometer shelterbelts spaced 200 meters apart would protect roughly forty square kilometers of farmland. More than 1,000 kilometers of shelterbelts have been planted in Nigeria over the past fifteen years, and another 1,700 kilometers are now planned.

The second type of investment involves the planting of trees on farm boundaries and near dwellings by the farmers themselves. Generally fifteen to twenty trees per hectare of farmland are considered desirable, and the species may be chosen because it provides fruit or fodder as well as shelter and fuel. The trees are less densely planted than in shelterbelts but are more uniformly distributed over the farmlands. They can have significant ecological benefits if—this is a major qualification—a large proportion of the farmers adopt the practice. The costs to the government of this type of investment are significantly lower, because the costs of planting and protecting the trees are borne by the farmers themselves; the forestry services are involved only in establishing nurseries and in providing agricultural extension services for farmers. If the farmers' response is good, afforestation over very large areas (much larger than with shelterbelts) becomes financially and institutionally feasible. The uncertainties about farmers' response are significant, however, and the risks much greater than with shelterbelts. In this chapter, the shelterbelt programs are distinguished from tree planting by farmers, which is termed "farm forestry."[1]

Costs and benefits can be estimated in seven steps: (1) determining gross and net farm income, (2) determining the growth of agricultural productivity, (3) determining the rise in gross farm income as a result of protecting the environment, (4) calculating the rate of change in soil fertility, (5) calculating the value of wood per hectare farmed, (6) determining the costs of a project, and (7) computing the value of the land area occupied by trees.

Gross and Net Farm Incomes

Benefits can each be measured in terms of the effects of trees on soil fertility and thus on farm budgets. The starting point is the familiar farm-budget exercise for determining the value of crops and livestock (see, for example, Gittinger 1982). In northern Nigeria, the main crops are sorghum and millet. The combined gross value of these crops and livestock was about US$200 per hectare per year in border prices (1986).[2] Farm costs were about 85 percent of the gross value of output. The model uses the following notation and baseline parameters:

Gross farm income in year t, Y_t = US$200 per hectare for $t = 0$

Farm costs = 85 percent of Y_0

Growth of Agricultural Productivity

The above values of gross farm output and costs relate to traditional agriculture in the region. But it is necessary to allow for technical progress, which may raise the yields of agricultural and livestock activities over time. The analysis discounts the cost and benefit streams at a rate equal to the opportunity cost of capital. It would therefore be inappropriate to ignore the possibilities of technical progress. The emerging threat of soil erosion is not only that it makes traditional food production activities unsustainable, but also that it may preclude the application of technological improvements in the future. Put another way, it would be inconsistent to discount future benefits at a rate of, say, 10 percent and then argue that there are no prospects for growth of such an economically important activity as agriculture; one must either lower the discount rate when assuming no growth in agriculture or accept the higher discount rate and the possibilities that agriculture will grow.

In the past, Nigerian agriculture has not developed well, and attempts to improve husbandry and technologies have not met with success; according to the World Bank (1986), per capita food production had actually declined by five percentage points in the preceding decade. It is now generally agreed, however, that poor growth in agriculture has been a consequence of policies that deliberately favored industry, such as protection against industrial imports and direct and heavily subsidized state investment in industry. The oil boom years also raised the exchange rate and made food imports cheaper. Policies are changing, however, and the prospects for growth in agriculture under more favorable prices and patterns of public investment cannot be neglected. In the present analysis, three assumptions were made: (1) that the rate of growth (g) of real gross farm incomes is 3 percent per year; (2) that the rate of growth (g) of farm costs is also 3 percent per year; and (3) that the year when farm incomes begin to grow is 15 years after $t = 0$. The third assumption is to allow for lags in the efforts of the agricultural research and extension agencies and for the incentives of more favorable prices.

Rise in Gross Farm Incomes
as a Result of Environmental Protection

Consider shelterbelts first. Numerous field experiments in more than twenty countries during the past seventy years have shown that properly oriented and designed shelterbelts have significant effects on crop yields.[3] The range of effects varies widely depending on local climate, terrain, soils, and crop conditions, but are generally in the range 10 to 30 percent, with measurements of 50 percent or more sometimes being reported. The main benefits arise from reductions in surface wind velocities, reductions in crop damage, and, an especially important factor, reduced surface evaporation and increased moisture in the soil. Further, there is some evidence to suggest that the effects on crop yields are greater in drought years than in wet and average years because of the greater importance of soil moisture retention in these conditions. As a rule of thumb, foresters usually work with a net effect on yields of 15 to 25 percent (though of course this is no substitute for local measurements). This implies an increase in gross farm income without a significant increase in farm costs (apart from the relatively small effect of harvesting a heavier crop).

For farm forestry, data from field experiments are more scarce, and scientific research is needed. Most reports on the subject suggest that the ecological effects of planting trees are significant (see, for example, National Academy Press 1984). For the purposes of analysis, and assuming a good farmer response, a 5 to 10 percent effect on yields is considered conservative.

In sum, the model makes the following assumptions for the increase in gross farm income. Shelterbelts can raise gross farm income by 15 to 25 percent, and farm forestry can raise gross farm income by 5 to 10 percent. The economic effects of shelterbelts will be felt from seven to ten years after planting, while farm forestry projects take eight to thirteen years to affect farm income. (A gradual build-up from zero to full effect is assumed in these percentages.) As noted, these effects raise the gross value of farm output without significantly raising farm costs.

Rates of Change in Soil Fertility

Much of the concern over losses in tree stocks focuses on losses in soil fertility. Afforestation programs may not only prevent the further decline of the soil, keeping land from being lost to production, but may also restore and even improve fertility. As in the analysis of ecological problems in developing countries, scientific data on the actual changes taking place in soil fertility are scarce. Possibly this is because of the large areas involved and the differences between areas, so that what applies in one area does not apply in another. Again, there is no good substitute for local scientific research. Given the visible evidence of gulley erosion, abandoned farmlands, and the inferred effects of wind erosion after a decline in tree cover, there seems little doubt about the threat of losses in soil fertility. The main question to be resolved concerns the rate at which the loss is taking place. For the present study, the assumptions used were (1) that the rate of decline of soil fertility is from 0 to 2 percent per year; and (2) that the decline is stemmed after eight years with an afforestation project. Note that a decline in soil fertility reduces the gross value of the crop and livestock activities but not the farm costs. Land is taken out of production when the net value of farm output becomes negative.

Value of Wood Produced per Hectare Farmed

The value of the wood produced can be calculated from market values of wood and fruit and from the yield data supplied by foresters. For the shelterbelts the figure (in border prices) turns out to be US$22 per hectare per year once trees are established, and US$7 per hectare per year for farm forestry. Both of these figures are net of labor costs of harvesting, but not of investment, which is treated separately under project costs; the figures can thus be added to the value of net farm income per hectare.

Project Costs

Project costs are determined from calculations of physical input requirements made by foresters and agronomists. For the shelterbelts, the main costs are those of fencing and of the labor to plant the seedlings. Investment costs are about US$150 per hectare of farmland protected. In addition, compensation has to be paid to the farmers whose land is used for the shelterbelts; these can be significant, amounting to one third of the financial costs of shelterbelts in Nigeria. Compensation costs are estimated separately, by determining the economic value of the area covred by trees (see below). For farm forestry, the investment costs are much lower, about US$40 per hectare. The main costs are for the establishment of a nursery network, seedling distribution facilities, the inclusion of advice on farm forestry to broaden the extension message (including some additional training for the extension workers), and the farmers' labor involved in planting and protecting the seedlings. The farmers' labor costs amount to about 40 percent of the total costs of farm forestry. Allowing for the costs of compensating farmers, the financial costs of shelterbelts are about eight times those of farm forestry (at least in the present case) per hectare of farmland protected. As noted earlier, because farm forestry has fewer budgetary and labor constraints, it offers prospects of a major increase in afforestation rates if the problems of obtaining farmers' cooperation.

Area Occupied by Trees

The imputed annual value of land occupied by trees is taken to be the net annual income of the cropping and livestock activities that would utilize the land without the investments in afforestation. This annual value can be estimated directly from gross and net farm income, growth of agricultural productivity, and rates of change in soil fertility; the value declines over time with losses in soil fertility. The areas taken up by the trees are estimated to be 12 percent of the farming area for shelterbelts, and 2 percent for farm forestry.[4]

To sum up, the net benefits of the investments equal the present value of the net farm income with the investments, plus the net present value of the wood and fruit produced, less the present value of the net farm income without the investments, less the present value of the costs of the investments. The change in the net farm income is the most dominant feature. Without the investments, net farm income declines as soil fertility deteriorates; when it becomes zero, the land is abandoned. With the investments, the decline in soil fertility is slowed; then fertility is enhanced as wood and fruit are harvested; and eventually soil fertility rises as opportunities for technical progress in agricultural and livestock activities occur on soils that would not otherwise remain arable.

Results

Figure 10-1 summarizes the variations in farm income streams using the above model (1 naira = US$1 at the time of the study). The income

Figure 10-1. *Farm Incomes with and without a Shelterbelt Project: The Base Case*

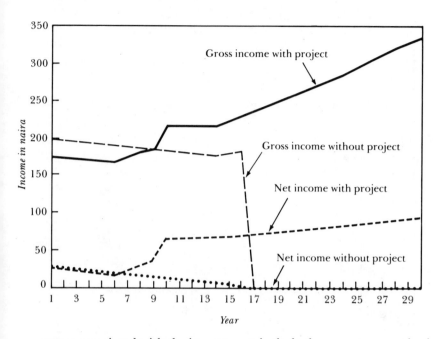

curves associated with the investments include the average annual value of the wood produced. The figure shows the case for shelterbelts, using the parameters noted earlier. (The farm forestry curves, which are not illustrated, follow a similar pattern). Gross income without the project is initially higher than in the case with the project because land is taken out of production to establish the shelterbelts. In both cases, gross and net farm incomes decline initially because of losses in soil fertility. With the project, the decline is eventually arrested and then reversed because of the factors noted previously, whereas without the project, gross and net farm incomes continue to decline until the net income becomes negative and land is taken out of production. Over a longer period, technical progress in agriculture (beginning in year 15) begins to exert a significant influence on gross and net farm incomes in the "with project" case.

Given the uncertainties in the ecological parameters and in the response of farmers to the farm forestry program, the calculations were repeated for a range of assumptions: lower and higher yields from the effects of shelterbelts and farm forestry, alternative rates of decline in soil fertility, and alternative cost assumptions. For example, the effects of a response from farmers can be studied in part by varying costs to allow for lags, for a more intensive extension effort to encourage farmers to plant trees, and for losses and replacement of seedlings.

Table 10-1. Results of Benefit-Cost Studies
(percent)

Case	Yield effect	Costs relative to base case	Rate of decline of soil fertility	Benefit-cost results		
				Net present value[a]	Benefit-cost ratio	Internal rate of return
Shelterbelts						
1. Base case	20	100	1	170	2.2	14.9
2. Low-yield, high-cost case	15	110	1	110	1.7	13.1
3. High-yield case	25	100	1	221	2.6	16.2
4. No erosion	20	100	0	108	1.8	13.5
5. More rapid erosion	20	100	2	109	1.8	13.6
6. Soil restored to initial condition, plus yield jump	20+	100	1	263	2.9	16.9
7. Wood benefits only	0	100	0	−95	0.3	4.7
Farm forestry						
1. Base case	10	100	1	129	4.5	19.1
2. Low case (no high case assumed)	5	150[b]	1	70	2.3	14.5
3. No erosion	10	100	0	75	2.9	16.6
4. More rapid erosion	10	100	2	60	2.5	15.5
5. Soil restored to initial condition, plus yield jump	10+	100	1	203	6.1	21.8
6. Wood and fruit benefits only	0	100	0	−14	0.6	7.4

a. In naira per hectare farmed. A 10 percent discount rate was used.
b. This increase corresponds to a three- to four-year lag in farmer response, plus a 10 percent cost increase.

Table 10-1 summarizes the results. The prospective rates of return to afforestation projects seem extraordinarily high even with quite conservative assumptions as to their ecological benefits and notwithstanding the long lags associated with the programs and with tree growth. The wood benefits alone are quite modest (though not trivial), as one would expect in arid conditions. Once one allows for the benefits of protecting the soils from erosion and for actually enhancing soil fertility, however, the rates of return, net present value, and cost-benefit ratios take the very high values shown.

One seeming anomaly in the results is that the benefits are lower when the rates of erosion rise from 1 to 2 percent per year (compare cases 1 and 5 in table 10-1 for shelterbelts and cases 1, 3, and 4 for farm forestry). The reason is that, at higher rates of erosion, the value of enhancing the fertility of a yet more degraded soil is much less. For example, a 20 percent enhancement of yields when soil fertility has already been degraded by 20 percent is much less valuable than when it has been degraded, say, by 10 percent. This raises both a well-known implication for policy and a problem for research. The policy implication is that it is generally advisable to protect the better soils before the threat of rapid degradation is serious; that is, to implement the projects sooner rather than later. The question for research is whether tree planting programs arrest the decline in fertility and gradually raise yields (by 15 to 25 percent in the case of shelterbelts) relative to the degraded condition; or whether they arrest the decline, restore the soil to its initial condition, and then gradually raise yields. Some foresters believe that the latter may apply—in which case the rewards of afforestation would be appreciable indeed (see case 6 for shelterbelts and case 5 for farm belts).

Conclusions and Implications for Further Research

If afforestation projects are analyzed only in terms of their capacity to produce wood and other tree products, their returns in arid and semi-arid areas are generally low. But in terms of arresting declines in—and then actually enhancing—soil fertility, the benefits may be appreciable.

Both socioeconomic and scientific uncertainties are encountered in this type of analysis in many countries. On the socioeconomic side, clarifications are needed in regard to why such beneficial assets—tree stocks on farmlands—are depleted so much that the fertility of soil is threatened, and also why farmers so often ignore the seemingly beneficial and low-cost investments required for farm forestry. Some countries, such as India and China, have launched successful programs (see Burley 1982; Chandler and Spurgeon 1979; Wiersum 1984; and Hintz and Brandle 1984), but there are less successful cases too—otherwise the problem of soil erosion would be less severe than is now reported. Hypotheses abound: the commons problem, property rights, downwind and downstream (external) effects, the biases against agriculture and rural areas in economic policy, the propensity of low-income groups to discount the future highly, mechanization, and even ignorance are all cited as partial explanations. There is an obvious need for research by the social scientists to shed light on these matters and, of equal importance, to draw out the normative implications of their findings.[5]

Another socioeconomic question worth addressing is whether the price mechanism can be made to work better, for example, by taxing farmers for not having trees or offering subsidies for having them. Alternatively, the policy could be an aspect of forestry law. Such an approach would be fraught with administrative difficulties and the dangers of funds being misappropriated, but the idea is worth analysis.

On the scientific side, the research agenda—especially in Africa—is equally huge. The rates of soil erosion, changes in soil fertility, and changes in tree stocks and the MAI are not known reliably, if at all, for large areas where an ecological threat is perceived. Other problems relate to the measurement of the yield effects of afforestation programs. Much research has been done in Asia, Europe, and North America on the yield effects of shelterbelts and windbreaks, but only scattered data are available for African countries. There is obviously no good substitute for local measurement inasmuch as ecosystems vary greatly as do the responses of soils to afforestation. Further, more is known about the effects of shelterbelts on soil fertility than about the effects of more dispersed plantings of trees by farmers. Yet farmer participation offers immense possibilities for a major acceleration of afforestation programs.

Despite the uncertainties arising from the lack of scientific and social research, policies and investments that address the ecological problems noted above should not be delayed, pending the outcome of further research. Procrastination is often politically convenient for policymakers, but it is not necessarily the right decision. Much can be recommended and implemented in the face of significant uncertainty. Uncertainty raises the risks of investments and of any attendant policy decisions, but the risks are still worth taking; indeed, it is likely that the ecological risks of procrastination far outweigh the financial and economic risks of investments that might not succeed in the first or even second attempt.

Less can be accomplished in the presence of the uncertainties mentioned, and special features need to be incorporated in the design of policies and investments. In particular, and perhaps obviously, a phased approach is needed. The first phase should include (1) an investment program sufficiently large to develop and demonstrate the approach; and (2) a significant component to finance social and scientific research, to monitor the progress of the investments, to monitor developments outside the project (such as the ecological changes taking place, the rate at which land is being abandoned, and social change) and more generally to lay the basis for a long-term plan.

Notes

1. Foresters often include shelterbelts under the heading "farm forestry." There is a distinction between the two approaches. For a further discussion of shelterbelts and farm forestry, see Burley (1982); Wiersum (1984); and the National Academy Press (1984).

2. Details for calculating these figures are available in Anderson 1987. The figures are based on an analysis of agricultural development projects in the region.

3. New field data and reviews of evidence from field trials and wind tunnel experiments are reported in Brandle and Hintz (1988).

4. It has since been suggested to me that for farm forestry to have a significant effect on wind velocities, the 2 percent figure is too low; 3 to 5 percent may be more appropriate once the trees have reached maturity.

5. Megan Vaughan of Nuffield College, Oxford, England, has kindly drawn to my attention a bibliography for social sciences, Seeley (1985), that contains more than a thousand references.

References

Anderson, Dennis. 1986. "Declining Tree Stocks in Africa." *World Development,* 14(7):853–63.

———. 1987. *The Economics of Afforestation: A Case Study in Africa.* World Bank Occasional Paper 1 (New Series). Baltimore, Md.: Johns Hopkins University Press.

Brandle, James R., and David L. Hintz, editors. 1988. "Proceedings of an International Symposium on Windbreak Technology, June 23–27, 1986." *Agriculture, Ecosystems, and Environment* 22–23 (August): 1–598.

Burley, Jeffrey. 1982. *Obstacles to Tree Planting in the Arid and Semi-Arid Lands: Comparative Case Study from Indian and Kenya.* Tokyo: United Nations University.

Chandler, Trevor, and David Spurgeon, editors. 1979. *Conference on International Cooperation in Agro-forestry.* Nairobi: DSE (German Foundation for International Development) and the International Development Council for Research on Agroforestry.

Gittinger, J. Price. 1982. *Economic Analysis of Agricultural Projects.* 2d ed. Baltimore, Md.: Johns Hopkins University Press.

National Academy Press. 1984. *Agro-Forestry in West African Sahel.* Washington, D.C.

Seeley, J. A. 1985. *Conservation in Sub-Saharan Africa: An Introductory Bibliography for the Social Sciences.* Cambridge African Monographs 5. Cambridge University, Cambridge, U.K.

Wiersum, K. F., editor. 1984. *Strategies and Designs for Afforestation, Reforestation and Tree Planting*. Netherlands: Pudoc Wageningen.

World Bank. 1986. *World Development Report 1986*. New York: Oxford University Press.

11

Multilevel Resource Analysis and Management: The Case of Watersheds

John A. Dixon

The management of environmental resources to promote economic development is of growing concern around the world. A country's natural resource base is important to the provision of a wide range of products. Agriculture, forestry, and fisheries are all directly dependent on the wise management of resources for their continued productivity. Industrial development and urban growth are indirectly dependent on many of the same natural resources.

Initially, efforts to improve environmental management focused on large projects—the development of industry, infrastructure, and irrigation, for example—that created well-defined problems. Improvements in the engineering and design of some such projects did indeed mitigate their negative effects on the environment. But in many cases, especially in developing countries, the chief cause of environmental problems is the cumulative effect of thousands or millions of individual micro-decisionmakers. For example, a large dam and reservoir project may be properly designed to control soil erosion during the construction phase. Over time, however, the actions of many individual farmers, both above and below the dam site, may have a much larger aggregate impact on soil erosion and sediment loads in the river.

Another example of the environmental effects of individuals is the air pollution that is damaging the white marble of the Taj Mahal in India. An initial environmental analysis to determine the cause focused on a large refinery project in the city of Mathura, some distance from Agra. Closer examination, however, showed that although emissions from the Mathura refinery could cause some of the measured air pollution affecting the Taj, a much more important source of pollutants were the hundreds of small foundries and workshops in the city of Agra. The refinery was an obvious target for pollution control, but the foundries and other small industries were a much more important source of total pollution. The problem is that technology and regulations exist to control the big project, the refinery, but the small-scale polluters nearby are harder to regulate or control.

In this and many similar cases, a paradox exists. Traditionally, efforts to protect the environment have focused on big projects (such as those commonly funded by international development banks and national governments) and on appropriate procedures for making them environmentally sound. Although these efforts have been successful, it may be a case of "winning the battle yet losing the war." Unfortunately, it has been much more difficult to reach and influence the small-scale individual resource user (the farmer, fisherman, pastoralist, or small-scale industrialist). Yet it is with these individuals that the war will be won or lost.

A new and more comprehensive approach is required, therefore, to improve resource management and make it truly effective. Such an approach must utilize multilevel analysis and management in order to account for the consequences of both the large development project and the individual resource user.

The growing realization of the important role of macro-level policies for improved environmental resource management is reflected in other chapters of this volume. As Warford (chapter 2) indicated, policies are designed to influence individual decisions about resource use by means of market signals that extend throughout the economy. Sometimes resource mismanagement is due to problems with common property, as in the case of overexploitation of fisheries or excessive air pollution. More often, however, mismanagement is due to market distortions caused by policies that send the wrong signal. Prices may not reflect the true opportunity cost of using a resource, which can lead to overuse. Repetto (chapter 6) cites examples of the negative impacts on resource management of policies related to pesticides, irrigation water, and crop prices.

Watersheds as Multilevel Systems

Watersheds present an interesting example of this multilevel problem. Watersheds are geographically discrete land and water systems that are composed of physical, economic, and social systems. They frequently are the focus of major development projects—a dam, an agricultural development, an industrial facility—and care is usually taken to ensure that these projects are developed in an environmentally sound way. At the same time, watersheds have proved particularly challenging to manage successfully, in part because of the complexities of dealing with different resources and groups of users in large, diverse areas. (The term watershed is used here to refer to a smaller spatial unit than a river basin. Some major river basins include large parts of entire countries and present special analytical problems; such major systems are not the topic of this chapter.)

It is widely perceived that there are serious problems in many watersheds, especially in developing countries. The following problems are symptomatic of resource mismanagement.

- Large-scale, massive deforestation and encroachment into "protected" areas in upper watersheds has been documented in many parts of the world, especially Asia and Latin America. In the Philippines, for example, more than 14 million people, or a quarter of the entire population, were living in upland areas in 1980 (Cruz 1986). This large and growing population places a heavy strain on the land and water resources of the uplands.
- Expansion of permanent and shifting cultivation into upper watershed areas continues. These areas are frequently unsuitable for long-term cultivation because of slope, soil characteristics, or the cultivation practices used. Once the soil loses its fertility, cultivators move on to other lands, leaving degraded and erodable soils behind.
- Soil erosion, sedimentation, and changed water flows are the result, often with very harmful effects on downstream resource users. For example, increased sedimentation means that the stream of benefits from many dams will be both reduced in magnitude and shortened. In one study (reported in Sfeir-Younis, 1986b), it was estimated that one-third of the capacity of all the major dams built since 1940 will be lost by the year 2000 if 2 percent of live capacity is lost each year through sedimentation. Much higher loss rates have been reported in Asia and elsewhere. Premature filling of reservoirs by sediment imposes large economic costs on society. Not only is invested capital made worthless and the flow of goods and services sharply reduced, but good sites for dam and reservoir construction are permanently lost, foreclosing future options. In a recent study, Mahmood (1987) estimated that the annual worldwide cost to replace lost reservoir capacity was US$6 billion. Part of this lost capacity is due to natural erosion and sedimentation, but a sizable amount is due to human actions.
- Changed patterns of water flow may cause flooding in the lowlands and changes in water quality. Although there is some dispute about the cause-and-effect relationship, it is clear that there is a connection between watershed management and water flow (Hamilton, 1983, 1985).
- Decreased fish production in both stream and coastal fisheries is one result of increased sediment loads in river systems.

It is easier to identify the effects and the causes of watershed mismanagement than it is to identify remedies. In one sense the problems are
the result of what has been called "the tyranny of small decisions." Negative effects result from attempts by countless individuals to use and manage their resources in a way that maximizes their own welfare. And yet
the outcome does not maximize total welfare.

In such a situation it may be myopic to analyze the problem
atomistically, on the basis of individual resource users. A more aggregate level is needed, but one that is less than national in scope. The watershed has considerable merit as a physical and economic unit of
analysis, and its potential and limitations as a resource management unit
are explored here.

Given the interaction of gravity and topography, most of the important physical effects and ecosystem interactions are contained within a
watershed system. If a watershed extends to the coast and adjacent
coastal waters (a convenient downstream boundary), a careful physical
and economic accounting will include most relevant effects. This view
has frequently been cited as a rationale for watershed or river basin planning (Schramm 1980). For example, when a group of researchers at the
East-West Center began thinking about watershed management, they
initially believed that a watershed as a planning unit included most
cause-and-effect relationships that were important, and that economic
analysis of this unit could accommodate most of these interactions in a
theoretical framework for improved resource management. It is now
clear that this view is incomplete. Actions or policies that are external to
the watershed (trade or price policies, for example) may have important
effects within the watershed system, and an economic analysis, even an
extended social welfare analysis, can capture only part of the important
interactions within the watershed (see Easter, Dixon, and Hufschmidt
1986). The proper management of watersheds therefore requires that
they be considered as multilevel systems including land, water, and
sociopolitical components.

The combination of natural systems and social systems that coexist in
a watershed is illustrated in figure 11-1. A typical watershed is an overlay
of natural and social systems. The natural system is defined by the land
and water base and the social system determines how these resources are
used. Government policies, as an extension of social organizations and
institutions, influence patterns of resource management.

The Management Challenge

Watersheds are seen as a resource management problem largely because
traditional approaches have failed to produce the desired results. Careful attention to individual projects in watersheds is not sufficient. The

Figure 11-1.   *Schematic of Natural and Social Systems of a Watershed*

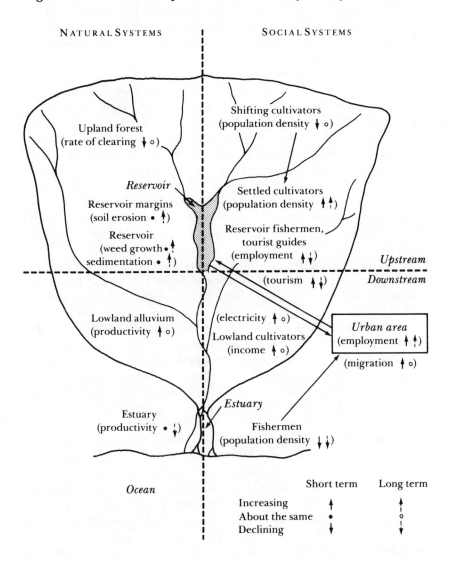

Source: Easter, Dixon, and Hufschmidt (1986), p. 8.

negative effects noted earlier are the results of complex interactions; solutions are not always clear. A multilevel integrated analysis is needed to identify linkages and design policies and programs that will then be carried out by existing sectoral agencies or local administrators and managers. Economic analysis has a key role to play in examining the interactions of different parts of the watershed system—land and water resources, people, organizations, and institutions. Economic analysis requires that most variables be expressed in monetary terms. To do this in the context of management of watershed resources requires several steps, beginning with an understanding of the physical environment.

Considerable literature exists on the physical interactions found in a watershed (for an overview, see Hamilton and Pearce 1986). This information is needed to identify the effects of human use or natural events in the watershed. Forestry activities, agricultural development, road and infrastructure construction, and urban developments are all activities found in major watersheds. Each activity creates certain direct and indirect effects.

Logging, for example, produces various wood products but also has a major impact on the rate of soil erosion, depending on the harvesting technique used and how access roads are constructed. The complex ecological linkages associated with deforestation are discussed by Myers in chapter 5. He uses deforestation as an illustration of the interdependency of nature and humans. Myers notes that misuse and overuse of a natural resource such as forests "generates a backlash effect on other natural resources, including soils, water, hydropower potential, fish stocks, and gene reservoirs." These effects in turn harm the welfare of present and future generations. Similarly, the construction of a dam often attracts increased settlement and resource exploitation in the areas above the construction site—the unintended result of access roads, which open up previously inaccessible terrain. Even if new settlement is forbidden by law, such laws are frequently ignored.

After the various effects have been identified, the next step is the quantification of these effects to the extent possible. Quantification in physical terms is followed by valuation, the estimation of monetary values for those effects. Not all effects can be monetized; some effects may have to be included in qualitative terms (for example, aesthetic concerns, genetic diversity, and historical or cultural significance). Although economists and others disagree about the extent to which monetary valuation can be used for many environmental goods and services, a great deal of recent work has focused on extending the range of effects that can be monetized (chapter 5 in this volume; Dixon and others 1988; Dixon and Hufschmidt 1986; Hufschmidt and others 1983; Pearce 1987; and Sinden and Worrell 1979). For example, Pearce (1987) estimates the value of dung used for domestic fuel in terms of the value of forgone agricultural production if the dung had been used as fertilizer (see also Newcombe, chapter 8). Other examples include the relationship between sediment loads and hydropower generation, or between turbidity and the production of fish in reservoirs. A common management problem in the upper watershed is that excessive soil erosion from agricultural activities causes increased sedimentation in a downstream reservoir. The costs of soil erosion can be measured in various ways—by the cost of reduced crop yields in the upland areas, in-stream problems of water quality or quantity, and effects on reservoirs, such as a reduction in the capacity to generate hydropower, the degradation of potable water, a decrease in the availability of irrigation water, and increased dredging or desiltation. All of these costs can be estimated.

In this volume, Pearce and Markandya (chapter 4) define the concept of marginal opportunity cost (MOC) as an organizing concept for economic analysis of resource management options. Very similar to marginal social costs, MOC includes the three types of costs associated with resource use: direct production or extraction costs (marginal direct cost, MDC), external environmental costs (marginal external cost, MEC), and the costs to future generations of using finite resources today (marginal user costs, MUC). The techniques for identifying the first set of costs (MDC) are well developed, and much work has recently been done on identification of the second set of costs (MEC). The third set, marginal user costs, is much more difficult to value and Pearce and Markandya give a simplified example of its use. Still, recent research has shown that with careful work a great deal can be done to include the external environmental costs and even the future opportunity costs of resource use.

The next step, the actual evaluation of a project or several alternative projects, is usually done in a benefit-cost or cost-effectiveness framework. The methodology for this is well established, and numerous authors, including those from the World Bank, have addressed this problem (see, for example, Gittinger 1982; Hufschmidt and others 1983; Little and Mirrlees 1969; Ray 1984; Squire and van der Tak 1975; and Dasgupta, Marglin, and Sen 1972). Sometimes the economic analysis is confined to direct project costs and benefits. In other cases an "extended" approach is used and various environmental effects and economic externalities are included in the analysis; that is, both MDC and MEC are included. In either case the underlying basis of welfare economics and the analytical tools used are similar. The decision criterion employed to evaluate alternative projects may be an internal rate of return (IRR), a benefit-cost ratio (BCR), or a netpresent value (NPV). All three criteria use the same basic input data and their application depends on the constraints and situation in each project. Guidance on which criterion to use has been provided in several studies (Dixon and Hufschmidt 1986; Gittinger 1982).

This overall approach is well developed and widely used. It can be applied to large and small projects, although large, discrete projects receive the most attention. Nevertheless, many watershed management projects promising a high IRR, BCR, or NPV are only partially successful or even outright failures.

Reasons for Failure

Several possible reasons for the disappointing performance of many land and water resource management projects and programs in watersheds are discussed here. Although not all reasons are applicable to all areas, some are found in most watersheds.

The Link between Cause and Effect Is So Far Removed in Time or Space That Social and Private Interests Diverge

In watersheds the link between cause and effect may be broken by space or time. Space refers to the classic concept of spatial externalities—one person's actions create an external effect (good or bad) on someone else who is spatially separate and not part of the decisionmaking process. When the link between cause and effect is spatially distant, the chances that proper land or water resource decisions will be made are reduced. In this case, "proper" usually means those actions that produce a result that is economically efficient when measured by social criteria. For example, a problem arises when the individual farmer measures his own losses from reduced productivity (and these losses may be small or zero) and compares them with the out-of-pocket costs of reducing erosion. Even though the appropriate technology is known (for example, terracing, contour plowing, agroforestry, or mulching), the farmer frequently decides that the benefits of preventing further erosion are less than the costs of taking the necessary measures, and therefore no action is taken. Conversely, if perceived benefits exceed costs the required actions would be taken: for example, conservation of upland areas would make good economic sense to the farmer if the farmer somehow received the benefits of a reduction in reservoir sedimentation. When the link between cause and effect is restored, more appropriate decisions are made.

Time is the other potential gap between cause and effect. Subsistence farmers see erosion occurring, but, because of a high personal discount rate, may be unwilling to take actions that yield significant benefits five or ten years in the future. The formal economic analysis may show a positive net present value of an activity when analyzed using a 8 or 10 percent discount rate; the farmer's own analysis may show a negative NPV using his own implicit 25 percent discount rate.

When the cause-and-effect link is broken for spatial or temporal reasons, it is necessary to use policy measures such as taxes, subsidies, or regulation to create incentives for improved management practices in land and water resource use. This point will be discussed later.

Social Dichotomies Lead to Problems in Resource Management

Sometimes watershed problems are a direct result of social problems. In many countries different ethnic or social groups inhabit the upper watersheds and the lowland areas. Upper watersheds are frequently marginal areas inhabited by politically or socially marginal people. Upland groups are often considered to be a "problem"—they are seen as illegal settlers and the cause of water-related problems downstream, such as flooding or deteriorating water quality. In turn, the people in the upper watershed consider themselves to be an oppressed minority and, not surprisingly, have little interest in taking actions to prevent damages downstream. These feelings lead to core-periphery problems that generate mistrust, alienation, and open hostility.

Lovelace and Rambo (1986) examined the general interaction between upland, lowland, and urban ecosystems and social systems in a watershed, while Dani (1986) explored the specific situation in the Hindu Kush-Himalayan region. Certain themes were common to both analyses. Within a watershed both sides may have valid arguments: the lowland-based rulers are concerned about illegal settlement on state lands in the upper watershed. The settlers perceive an indifferent or hostile government trying to prevent them from making a modest living for their families, all for the sake of electricity or water benefits that they do not receive. Frequently, the groups become polarized and it becomes very difficult to implement rational management programs for land and water resources.

Proposed Measures for Resource Management Are Clearly Unprofitable and Uneconomic

When measured by any standard criterion—either that of the government or the individual farmer—the proposed management actions may be clearly uneconomic. Sometimes a quick economic and financial analysis can clearly identify alternatives that should be avoided, that is, projects where costs greatly exceed any reasonable estimate of benefits, even at low or zero discount rates. The analysis is complicated when nonquantified and nonmonetized environmental benefits are thought to be relatively large, as frequently happens in arguments for the preservation of genetic diversity in tropical forests or other ecosystems. In these cases, conservationists and environmentalists are pitted against economists and government officials. The economic analysis may show costs greatly exceeding benefits, while preservationists feel strongly that the unquantified benefits easily justify project costs. There is no single correct answer; the ultimate decision is a political one whereby decisionmakers weigh the alternatives and decide how much preservation society wants or can afford. The role of the economist and of other scientists is to present the available information from their own perspective as clearly as possible.

Implementation Problems May Cause Project Shortcomings or Failure

Difficulties in implementation occur particularly when several different units must provide separate inputs in a closely coordinated fashion. Timing and spatial distribution are hard to coordinate, and separate budget cycles and organizational structures further complicate coordination. A classic implementation problem is the late arrival of key agricultural inputs such as credit, fertilizer, or seed after the appropriate time in the crop cycle has already passed. In the end, all of the required parts of the technical package were available in the field, but not in the right sequence or at the right time.

Frequently the bureaucratic structure is a hindrance to effective watershed management. For example, at the same time that the irrigation bureau is planning to divert a stream for supplemental irrigation, the utilities bureau may be planning to interrupt the flow with a dam, and the forest bureau may be granting a logging concession in the upper watershed that will lead to increased soil erosion and sedimentation in the reservoir. One possible response to these problems is to centralize all activities in one "super agency," but this is usually not a workable solution. The challenge is to work with existing, established organizations and to develop an implementation plan that builds on existing roles and reward structures.

Direct monitoring or management may be feasible for the small number of major projects but is usually not possible for the thousands of individual activities taking place in a watershed. To reach these small, individual decisionmakers, one must rely on the creation of proper incentives and the role of the market to send the correct signals.

A Philippine Case Study

The Upper Pampanga River Project is the Philippines' first large-scale, multipurpose water resources development project and centers around the Pantabangan dam and reservoir. The dam, completed in 1977, impounds approximately 3 billion (thousand million) cubic meters of water. With associated facilities the project cost more than 900 million pesos (approximately US$120 million). Upon completion the project was expected to irrigate 100,000 hectares of land in both wet and dry seasons, generation of 232 million kilowatts of electricity annually, and reduce flood damage to crops, livestock, and infrastructure (David 1987).

Unfortunately, the project has not reached these targets. The population has grown, in part because of rapid in-migration associated with construction activities. Most of the population is now dependent on agriculture for their livelihood. Much of the land development has been done in an unsustainable manner that has caused soil erosion and reservoir siltation. As Cruz (1988) stated, "soil erosion does not necessarily impose current costs on the private land user as long as the topsoil layers are not completely depleted. Since the upland farmer has no right to the land and therefore no stake in ensuring its long-term productivity, the potential gain by reducing the loss of soil nutrients cannot be captured by the farmers. It is therefore not surprising that upland farmers exploit the land until its productivity declines and then move on to a new plot."

As a result of these and other problems the original sedimentation rate estimated for the Pantabangan reservoir (20 tons per hectare per year) is only about one quarter of the estimated present amount (David 1987). Even with the enlargement of dead storage area, the service life of the reservoir has been decreased from the orginal estimate of 100 to 61 years; the magnitude of benefits from the stored water has also been decreased.

Estimates of on-site and off-site costs of soil erosion in the Pantabangan area have been calculated (Cruz 1988). The main on-site effects are related to loss of organic matter and nutrients; the main off-site effects are reservoir sedimentation and loss of hydropower, irrigation water, and flood protection.

In attempting to counteract these effects, the government has launched a watershed rehabilitation and management project aimed at the individual resource user. Progress has been slow and the challenge is large. In an analysis of the Pantabangan watershed, the National Irrigation Administration highlighted some of the major problems: pervasive rural poverty; rapid land conversion and accelerated soil erosion; inadequate government regulatory and management capability; ineffective land use planning; poor delivery of agricultural support services in the uplands; and critical gaps in available technology, knowledge, and manpower for proper resource conservation.

In addition, analysis of the Pantabangan problems indicates that a large amount of sediment is already in transport within the system and may not reach the reservoir for some time. Even if all on-farm erosion ended tomorrow, sedimentation of the reservoir would continue for many years.

The necessity of analyzing the reasons for resource degradation and designing appropriate management policies is critical. The answers, however, are not easy because a main target of any management plan, the individual farmer, is very poor and very hard to reach. Therefore, even though the causes and costs of mismanagement are known, it requires careful planning, financial resources, and political will to help solve the problem.

A Multilevel Approach to Resource Management

Management of watershed resources will benefit from an intermediate approach that builds on both micro-level analysis, to understand how and why individuals use resources in a certain way, and macro-level analysis in search of appropriate policy measures. The negative effects on the environment created by both large projects and individual resource users must be taken into account.

Large, discrete projects that affect an entire ecosystem because of their size can be both analyzed and managed at the level of the individual project. This is already being done. In contrast, the impacts of small-scale individual resource users, each of whom has only a small effect on the environment, must be analyzed at the individual level, but must be managed at the larger, policy level. To use Mies van der Rohe's famous architectural aphorism—less is more—in another context, the cumulative resource impact created by the actions of the small-scale resource user is what really matters for improved resource management. The megaprojects that receive much attention may have a proportionately smaller impact overall.

This dichotomy creates the need for multilevel analysis and management. At the analytical level, both large projects and the actions of individual farmers must be analyzed to understand what is happening. At the policy or management level, however, a different approach is necessary. Whereas the large project can be directly regulated or controlled, the individual resource user can be reached effectively only by the use of macro-level policies and incentive systems such as price policies, subsidies or taxes, or land reform. In the case of true subsistence farmers who are largely outside the market system, it may be necessary to use community-based group projects.

Regardless of the level of management, the following lessons can be drawn from examination of the watershed resource management experience.

First, an integrated, multidisciplinary approach is required. Given the complicated interactions between resources and people, any program designed from a single perspective (geological, economic, hydrologic, or political) is unlikely to be successfully implemented. A wide variety of expertise is required to identify how resources are being used, what the likely effects of that use will be, and why these actions are being taken.

Second, economic analysis can play a key role in such an integrated approach. Since many resource users are acting on their own perception of benefits and costs, it is essential to examine these different users to understand what macro-level policies or programs may be effective in improving their use of land and water resources. Economic analysis from both the individual or private perspective and from the public perspective is of key importance. It is necessary to examine the broader environment of economic policy and the impact of macro-policies on decisionmaking within the watershed (Sfeir-Younis 1986a, 1986b). Trade and pricing policies send signals through the market that affect resource use within a watershed. Even if such policies do not affect subsistence farmers in the uplands, they may help prevent the displacement of labor from lowland agriculture to the uplands.

Third, environmental externalities should be included, to the exent possible, in the analysis of social welfare. Recent advances in valuation methodology have made it increasingly possible to include monetary estimates of environmental effects. The monetary value of environmental benefits or costs, both at present and in the future, should be included in the economic analysis.

Fourth, it may be necessary to use different discount rates in private and public analyses. Economists traditionally insist that only one discount rate be used in the analysis. Because a public analysis has a larger spatial and temporal framework than that of an individual, however, it may be necessary to use one discount rate (usually a higher one) for the private analysis and another, lower discount rate for the public analysis of a single project. (It is incorrect to use different discount rates to analyze different projects and then compare the results.) Frequently the results of the two analyses will diverge, and then the government must devise appropriate management policies to bring about the desired result.

Fifth, multiple interests should be reflected in the policymaking. Just as several disciplines need to be involved in the analytical process, many different and sometimes conflicting interests need to be accommodated in policy formulation. Political, social, ethnic, and moral concerns are important and may dictate an alternative other than the economically preferred solution. This is sometimes referred to as the second-best solution. Economic criteria can be used to design the best solution, but there will frequently be valid reasons that make it unacceptable. Concerns for income distribution, minority rights, or political balance may dictate another approach. Economic analysis, carefully done, can indicate the tradeoffs and economic costs of second-best solutions. This information can be very useful for informed policymaking.

Sixth, integrated analysis does not mean integrated implementation. Integrated resource management programs can rarely be implemented by using existing institutional or organizational frameworks. The existing structures are usually organized on sectoral lines and may divide the watershed among different organizations. Although this obviously complicates program implementation, it is inadvisable to set up a new, integrated organization. Attempts to integrate over wide areas or across sectors have usually been unsuccessful. Organizations such as the Tennessee Valley Authority in the United States are rare exceptions. In some cases the use of integrated oversight committees to coordinate individual departmental work has been successful.

The lack of an integrated implementing authority is not a major problem if an integrated analysis is done to identify problems and desired solutions. Although program implementation is along existing organizational lines, integrated analysis can result in actions being taken by a certain ministry (for example, terracing in the uplands via the public works department) even when the costs clearly exceed the benefits to that ministry. In this example, the irrigation and hydropower authorities would reap large downstream benefits even though they do not pay for or implement the terracing program. The government may need to provide incentives to promote interministerial cooperation in order to facilitate project implementation.

Macro-level policies that rely on the market to send signals do not face the same organizational constraint. These policies are affected, however, by the extent to which some watershed residents are part of the market economy. The less market involvement, the less useful are such macro policies.

Seventh, government intervention often takes the form of fiscal and price policies. Because large projects are relatively easy to manage and influence, both direct intervention and macro-policies focus on the actions of the individual resource user. In many cases, the appropriate role for government is to provide the needed incentives and assistance for individuals or groups to undertake the desired actions themselves. These policies are the government's way of providing the link between cause and effect and of connecting costs and benefits over space and time. Some combination of taxes, subsidies, direct investment, or price changes may be needed.

In the case of upland terracing to protect lowland reservoirs, any of a number of policies could be used: regulations requiring farmer-built terraces, subsidies for farmer-built terraces, communal construction of terraces (with or without compensation), government construction of terraces, increased taxes on unterraced lands, reform of land tenure, dredging of lowland reservoirs, increased taxes on lowland water and power users to pay for the other options, changed input or output prices to encourage farmer-built terracing, and incentives for increased interministerial cooperation. No single policy or set of policies is correct or always applicable. Some policies require direct intervention to implement while others rely on market signals. The challenge for the analyst is to identify the policy alternatives that are acceptable given the social, cultural, and political contexts of each case and to determine the policy that can be implemented effectively. The economist examines the economic implications of alternatives, calculates the cost and benefits of each, and determines who "wins" and who "loses."Ultimately, it is the political decisionmaker who decides, but the integrated approach will help in this process as policymakers search for effective, efficient, and equitable solutions to land and water resources management.

References

Cruz, M. C. J. 1986. *Population Pressure and Migration: Implications for Upland Development.* Working Paper 86-02. Manila: Philippine Institute for Development Studies.

Cruz, Wilfrido. 1988. "Strategies for Sustainable Agricultural Development: Focus on Upland Agriculture." Paper presented at the Economic Development Institute/SEARCA (South-East Asian Regional Center for Graduate Study and Research in Agriculture) Regional Seminar on Land and Water Resources Management, Los Baños, Philippines, March 19–30.

Dani, A. A. 1986. "Annexation, Alienation and Underdevelopment of the Watershed Community in the Hindu Kush-Himalaya Region." In Easter, Dixon, and Hufschmidt (1986).

Dasgupta, Partha, Stephen Marglin, and Amartya Sen. 1972. *Guidelines for Project Evaluation.* New York: United Nations Industrial Development Organization.

David, W. P. 1987. *Hydrologic Validation of the Pantabangan Watershed Management and Erosion Control Projects.* Working Paper 87-03. Manila: Philippine Institute for Development Studies.

Dixon, John A., R. A. Carpenter, L. A. Fallon, P. B. Sherman, Suphachit Manipoke. 1988. *Economic Analysis of the Environmental Impacts of Development Projects.* London: Earthscan Publications.

Dixon, John A., and M. M. Hufschmidt, editors. 1986. *Economic Valuation Techniques for the Environment: A Case Study Workbook.* Baltimore, Md.: Johns Hopkins University Press.

Easter, K. W., John A. Dixon, and M. M. Hufschmidt, editors. 1986. *Watershed Resources Management: An Integrated Framework with Studies from Asia and the Pacific.* Boulder, Colo.: Westview Press.

Gittinger, J. Price. 1982. *Economic Analysis of Agricultural Projects.* 2d ed. Baltimore, Md.: Johns Hopkins University Press.

Hamilton, L. S. 1985. "Overcoming Myths about Soil and Water Impacts of Tropical Forest Land Uses." In *Soil Erosion and Conservation*, edited by S. A. El-Swaify, W. C. Moldenhauer, and A. Lo. Ankeny, Iowa: Soil Conservation Society of America.

Hamilton, L. S. (with P. N. King). 1983. *Tropical Forested Watersheds: Hydrologic and Soils Response to Major Uses or Conversions.* Boulder, Colo.: Westview Press.

Hamilton, L. S., and A. J. Pearce. 1986. "Physical Aspects of Watershed Management." In Easter, Dixon, and Hufschmidt (1986).

Hufschmidt, M. M., D. E. James, A. D. Meister, B. T. Bower, and J. A. Dixon. 1983. *Environment, Natural Systems and Development: An Economic Valuation Guide.* Baltimore, Md.: Johns Hopkins University Press.

Little, I. M. D., and J. A. Mirrlees. 1969. *Manual of Industrial Project Analysis in Developing Countries.* Vol. 2, *Social Cost-Benefit Analysis.* Paris: OECD.

Lovelace, G. W., and A. T. Rambo. 1986. "Behavioral and Social Dimensions." In Easter, Dixon, and Hufschmidt (1986).

Mahmood, K. 1987. *Reservoir Sedimentation: Impact, Extent and Mitigation.* World Bank Technical Paper 71. Washington, D.C.

Pearce, David. 1987. "Valuing Natural Resources and the Implications for Land and Water Management." *Resources Policy* (December):255–64.

Ray, Anandarup. 1984. *Cost-Benefit Analysis: Issues and Methodologies.* Baltimore, Md.: Johns Hopkins University Press.

Schramm, Gunter. 1980. "Integrated River Basin Planning in a Holistic Universe." *Natural Resources Journal* 20 (October).

Sfeir-Younis, Alfredo. 1986a. "Economic Policies and Watershed Management." In Easter, Dixon, and Hufschmidt (1986).

———. 1986b. "Soil Conservation in Developing Countries." Agriculture and Rural Development Department, World Bank, Washington, D.C. Processed.

Sinden, J. A., and A. C. Worrell. 1979. *Unpriced Values: Decisions without Market Prices.* New York: J. Wiley & Sons.

Squire, Lyn, and Herman G. van der Tak. 1975. *Economic Analysis of Projects.* Baltimore, Md.: Johns Hopkins University Press.

Index